AN ODYSSEY THROUGH THE BRAIN, BEHAVIOR AND THE MIND

AN ODYSSEY THROUGH THE BRAIN, BEHAVIOR AND THE MIND

by

C. H. Vanderwolf, Ph.D., DSC.
University of Western Ontario, London, Ontario Canada

KLUWER ACADEMIC PUBLISHERS
Boston / Dordrecht / London

Distributors for North, Central and South America:
Kluwer Academic Publishers
101 Philip Drive
Assinippi Park
Norwell, Massachusetts 02061 USA
Telephone (781) 871-6600
Fax (781) 681-9045
E-Mail < kluwer@wkap.com>

Distributors for all other countries:
Kluwer Academic Publishers Group
Post Office Box 322
3300 AH Dordrecht, THE NETHERLANDS
Telephone 31 786 576 000
Fax 31 786 576 254
E-Mail < services@wkap.nl>

Electronic Services < http://www.wkap.nl>

Library of Congress Cataloging-in-Publication Data

Vanderwolf, C. H.
An odyssey through the brain, behavior and the mind / by C.H. Vanderwolf.
p. cm.
Includes bibliographical references and index.
ISBN 978-1-4419-5335-3
1. Brain. 2. Neuropsychology. 3. Cognitive neuroscience. I. Title.

QP376 .V364 2002
153—dc21 2002040610

Printed on acid-free paper.
Printed in the United States of America

CONTENTS

PREFACE

Much of contemporary behavioral or cognitive neuroscience is concerned with discovering the neural basis of psychological processes such as attention, cognition, consciousness, perception, and memory. In sharp divergence from this field, this book can be regarded as an elaborate demonstration that the large scale features of brain electrical activity are related to sensory and motor processes in various ways but are not organised in accordance with conventional psychological concepts. It is argued that much of the traditional lore concerning the mind is based on prescientific philosophical assumptions and has little relevance to brain function.

The first ten chapters of this book give a personal account of how the various discoveries that gave rise to these views came to be made. This is followed by discussions of brain organization in relation to behavior, learning and memory, sleep and consciousness, and the general problem of the mind.

ACKNOWLEDGEMENTS

I thank the many students, post-doctoral fellows, colleagues and technical staff with whom I have been fortunate to work in the last 40 years on the topics discussed in this book: Glen Baker, Rick Beninger, Hank Bitel, Brian Bland, Francis Boon, George Buckton, Gyorgy Buzsaki, Peter Cain, Christian Caldji, Richard Cooley, Rick Cornwall, Bob Davis, Laszlo Detari, Clayton Dickson, Anthony Dobravec, Steve Donaghy, Hans Dringenberg, Lisa Eckel, Colin Ellard, Chris Engeland, Allan Fine, Adrian Gelb, Larry Gillespie, Mel Goodale, Mike Gutman, John Hamilton, I. Hanin, Eric Hargreaves, Greg Harvey, Bob Heale, Woody Heron, Tim Hoh, L.B. Hough, Nadia Jandali, Martin Kavaliers, Mary-Ellen Kelly, Steve Kendall, Bryan Kolb, Phil Kraemer, Ron Kramis, Stan Leung, S. Leventer, Derek McFabe, Michelle McLauchlin, Larissa Mead, Craig Milne, Janice Moore, Peter Moore, John Orphan, Peter Ossenkopp, Bruce Pappas, Barbara Peck, Debbie Penava, Karen Petersen, Danny Pulham, Angela Raithby, Barbara Robertson, Terry Robinson, Bob Sainsbury, Phil Servos, Debbie Stewart, Dwight Stewart, Angela Streather, Ken Strong, Henry Szechtman, Mary Vanderwart, David Wakarchuk, Neil Watson, Ian Whishaw, and Elaine Zibrowski. Without their help, this book could never have been written.

I also thank: (a) the Natural Sciences and Engineering Research Council of Canada for their unfailing support over many years, including support for the publication of this book; (b) the Department of Psychology of the University of Western Ontario for their support in many different ways; (c) Leon Surette for proof-reading the manuscript ; and (d) Daniella Chirila and Sarah Vanderwolf, who typed the manuscript for this book.

Chapter 1

A Preparation and a Beginning

In September, 1958, I arrived in Montreal to begin graduate work on the brain and behavior with professor D. O. Hebb in the Department of Psychology at McGill University. I was then 22 years old, a recent graduate of the University of Alberta [B.Sc. (Honours), 1958] and very much interested in the exciting findings in brain research that had been made at McGill in the preceding decade. In 1949 Hebb had published a book (*The organization of behavior*)[1] that inspired much subsequent research at McGill and elsewhere on the role of early experience in the development of normal adult brain microstructure and function, the role of a varied sensory input in the maintenance of normal brain function in adults, and on the change in neuronal function associated with learning and memory. W. Penfield was just at the end of his active career as a neurosurgeon at the Montreal Neurological Institute; he and H. Jasper had recently published "*Epilepsy and the functional anatomy of the human brain*"[2], a book which summarized their pioneering work on the effects of electrical stimulation of the brain in conscious humans. A great deal of neuroscientific work at McGill in this period centered on the role of the ascending reticular activating system in arousal, consciousness, and attention. Jasper had been the editor-in-chief of a symposium on the "*Reticular formation of the brain*"[3] which was published in 1958; an earlier symposium on "*Brain mechanisms and consciousness*"[4] held near Montreal, had been published in 1954. J. Olds and P. M. Milner had discovered (1954)[5] that a rat would rapidly and repeatedly press a lever which switched on electrical stimulation to certain parts of its own brain via a chronically implanted electrode. This seemed to provide a method for studying the localization of reward or pleasure in the brain. W. Scoville and B. Milner had just published a paper (1957)[6] suggesting that the hippocampal formation plays an essential role in the laying down of permanent memories. It seemed to many people at the time that the neural basis of attention, consciousness, reward, and memory would soon be well understood.

Exciting though all this was, it was necessary for me to get on with the practicalities of life. I found cheap lodgings in the McGill "student ghetto" and settled in for what became a nearly four year stint as a graduate student, supported by scholarships from the National Research Council (NRC). Hebb suggested I work on the behavioral functions of the medial thalamic nuclei, a part of which was believed at the time to be a rostral extension of the ascending reticular activating system. Since the facilities available for electrophysiological work in Hebb's lab were very limited, most of my experiments dealt with the behavioral effect of localized lesions in the medial thalamus in rats. In addition to this research, I attended a variety of lectures or formal courses in neuroanatomy, neurophysiology, pharmacology, and psychology. In December, 1961, I completed the requirements for a Ph.D. degree which was formally conferred the following spring.

During this period, I made an observation that developed into a major research program for me. One day, two fellow graduate students, Lloyd Gilden and Harry Cohen, were attempting (using the only working oscillograph available in Hebb's lab) to record the potentials elicitable in the striate cortex (visual cortex) by a bright flash of light. The rat was fully awake (the electrodes were chronically implanted) and, since it tended to wander away from the light source, Lloyd and Harry had attempted to immobilize it with a series of heavy wire hoops passing over the rat's body and driven into holes drilled in a board. Although this arrangement could not have been painful, the rat periodically made vigorous attempts to escape. I noticed that each of these episodes of struggling was associated with rhythmical potentials in the striate cortex and that these potentials disappeared the moment that the rat became immobile again. When I pointed this out to Lloyd and Harry, they watched for a time, agreed that it was so, but, regarding it as a thing of very little interest, went on with their work. This observation, nonetheless, found its way into an experiment I undertook with Dr. Bernard Woodburn Heron in the summer of 1962.

Woody Heron, as he was universally known, had been a post-doctoral fellow with Hebb and B. D. Burns (Physiology Department), doing research on the activity of single neurons in the striate (visual) cortex of the cat. Subsequently he had accepted a tenurable position in the Psychology Department at McMaster University in Hamilton, Ontario. As he and I had a beer together one day on one of his return visits to McGill, he began to question me rather vigorously about what I thought I had really learned as a result of my research on the medial thalamus. The main conclusion of my Ph.D. thesis had been that a part of the medial thalamus, probably the intralaminar nuclei, played a role in the control of movement, especially in the initiation of voluntary movement. Perhaps an electrical signal of some kind would be detectable in the intralaminar region just before a rat made a movement, such as running for example.

Subsequently, at Woody's invitation, I stayed with him and his wife Pauline in their apartment in Hamilton for about three weeks. We first trained rats to run from one compartment to another in a two compartment box (shuttle-box) whenever an auditory signal was presented. Subsequently, under anesthesia, we implanted electrodes in the medial thalamus, the hippocampus dorsal to the thalamus, and also in the neocortex and the brain stem reticular formation. Electrodes were placed in the hippocampus because I suspected that the rhythmical waves I had noticed in Lloyd's and Harry's experiment were being generated, not in the striate cortex, but in the underlying hippocampal formation. It had been shown in 1954[7] that electrical stimulation of the reticular formation produced rhythmical slow wave activity in the hippocampus via an ascending pathway through the septal nuclei.

When the rats had recovered from their surgery, Woody and I took recordings of brain electrical activity as the rats performed in the shuttle-box. We observed that rhythmical waves of 7-9 Hz appeared in both the medial thalamus and hippocampus just before the onset of walking or running and continued as long as the rats continued to move but disappeared promptly when movement ceased. These waves were not observed in the anterior part of the neocortex or in the brain stem reticular formation.

Although Woody and I concluded that these observations suggested some role of both the medial thalamus and the hippocampal formation in the control of motor activity[8] it must be admitted that many other interpretations were also possible. For example, L. Pickenhain and F. Klingberg, working at Karl-Marx University in Leipzig, in what was then called the German Democratic Republic, had observed a correlation between rhythmical hippocampal activity and certain types of motor activity at about the same time as Woody and I did, but they suggested that it indicated a relation to motivation[9]. It seemed clear to me that a far more extensive study of the problem was required. However, since I was to spend the next year as a post-doctoral fellow in California (with R. W. Sperry at the California Institute of Technology in Pasadena) and a second year in Switzerland (with K. Akert at the Brain Research Institute in Zurich) I could not return to the problem of the rhythmical waves until 1964 when I returned to McMaster University as an assistant professor with a small NRC research grant of my own and an opportunity to begin truly independent research.

The program of research that I had in mind was based on certain philosophical presuppositions. Most investigators in the brain-behavior field, both in 1964 and at present, assume that the ultimate problem for neuroscience is to provide an explanation of the human mind. More specifically, this is usually taken to mean discovering the neural basis of consciousness and its sub-processes such as perception, attention, memory, cognition, emotion, motivation, etc. I was very suspicious of this entire enterprise and gradually came to believe that the traditional categories of the mind did not provide a valid natural subdivision of different brain functions. During my career as a student, I had taken several laboratory courses in mammalian physiology, pharmacology and biochemistry and had been much attracted by the idea that a living animal can be regarded as an enormously complex machine whose operations are all potentially explicable in physical and chemical terms. Most of the brain, it seemed to me, was dedicated to the control of motor activity (behavior). If the entire forebrain and midbrain are surgically removed from an animal (leaving an island of the hypothalamus to permit operation of the pituitary gland), respiratory, cardiovascular, digestive, and excretory functions proceed almost normally but spontaneous behavior is abolished. Such a decerebrate animal can no longer walk about, feed itself, groom itself, seek shelter, avoid enemies, find a mate, or care for its young. Therefore, it must be that what the intact brain does is generate all of these (and more) varied behavioral performances. To say that a decerebrate animal has no behavior because it is unconscious seemed to me to be no explanation at all but merely a restatement of the problem in more obscure terms.

I had also a long-standing interest in what is now usually called animal behavior. This arose from my experience growing up on a farm in what was then a recently settled area in the mixed-wood boreal forest region of northern Alberta. As a teenager, spending most of my spare time hunting and trapping, I had learned a good deal about the woods and the creatures that lived there. Animal behavior was not a part of the curriculum when I was an undergraduate, but I eventually discovered the work of Niko Tinbergen[10]. Here was someone who not only possessed the type of intuitive knowledge of animals that hunters and farmers have but who also understood how to increase such knowledge by systematic scientific study. I hoped to contribute something to mammalian behavioral physiology.

I regret that I was not always very diplomatic in discussing brain-mind relations with colleagues. Leon Kamin, who was chairman of the Psychology Department during the latter part of the time I was at McMaster, took a kindly interest in me and tried to broaden my horizons by loaning me a book on "*Theories of motivation*". When he asked what I thought of the book a week or two later, I fear that I offended him by blurting out: "This stuff is like alchemy!"

Notes on Chapter 1

1. Hebb, D. O. (1949). *The Organization of Behavior.* New York: John Wiley and Sons, Inc.
2. Penfield, W., and Jasper, H. (1954). *Epilepsy and the Functional Anatomy of the Human Brain.* Boston: Little, Brown and Co.
3. Jasper, H., Proctor, L. D., Knighton, R. S., Noshay, W. C., and Costello, R. T. (eds.) (1958). *Reticular Formation of the Brain.* Boston: Little, Brown and Co.
4. Delafresnaye, J. F. (ed.) (1954). *Brain Mechanisms and Consciousness.* Oxford: Blackwell.
5. Olds, J., and Milner, P. (1954). Positive reinforcement produced by electrical stimulation of the septal area and other regions of rat brain. *Journal of Comparative and Physiological Psychology, 47*, 419-427.
6. Scoville, W. B., and Milner, B. (1957). Loss of recent memory after bilateral hippocampal lesions. *Journal of Neurology, Neurosurgery, and Psychiatry, 20*, 11-21.
7. Green, J.D., and Arduini, A. (1954). Hippocampal electrical activity in arousal. *Journal of Neurophysiology, 17,* 533-557.
8. Vanderwolf, C. H., and Heron, W. (1964). Electroencephalographic waves with voluntary movement: study in the rat. *Archives of Neurology, 11*, 379-384.
9. Pickenhain, L., and Klingberg, F. (1967). Hippocampal slow wave activity as a correlate of basic behavioral mechanisms in the rat. In W. R. Adey and T. Tokizane (eds.) *Structure and function of the limbic system; Progress in Brain Research, 27*, 218-227.
10. Tinbergen, N. (1951). *The study of instinct.* New York: Oxford University Press.

Chapter 2

Hippocampal Activity and Behavior

The problem I hoped to solve in 1964-65 was to determine the conditions under which rhythmical slow waves appeared in the hippocampus of the rat. The method adopted was to place rats with chronically implanted electrodes inside a box with a wood frame covered with a grounded copper screen (Faraday cage) to reduce extraneous electrical signals such as the ubiquitous 60 Hz line interference. One electrode, implanted in the skull, connected the rat to ground; others permitted the recording of slow electrical potentials from the hippocampus and the neocortex on an ink-writing oscillograph or polygraph via a light flexible set of leads made from phonograph pick-up wire. I sat close to the rat with a keyboard on my lap to allow me, by closing appropriate switches, to record on the polygraph record what the rat did.

The first attempts did not yield clear results. The hippocampal records were relatively low amplitude (200-300 μV) and although rhythmical slow waves were often present during gross movements such as walking, or struggling when a rat was picked up, they tended to be obscured by faster activity (mostly 15-50 Hz). The electrical currents associated with cerebral field potentials spread widely throughout the brain: hippocampal activity can be picked up by an electrode in the neocortex and large neocortical potentials can be picked up in the hippocampus. Eventually it became clear that a superior method of recording hippocampal slow waves consists of placing one electrode near the surface of the alveus or in stratum oriens (Figure 2-1) and a second electrode in the vicinity of the hippocampal fissure. Slow wave potentials at these two sites are of opposite phase so that when stratum oriens is negative with respect to an indifferent site

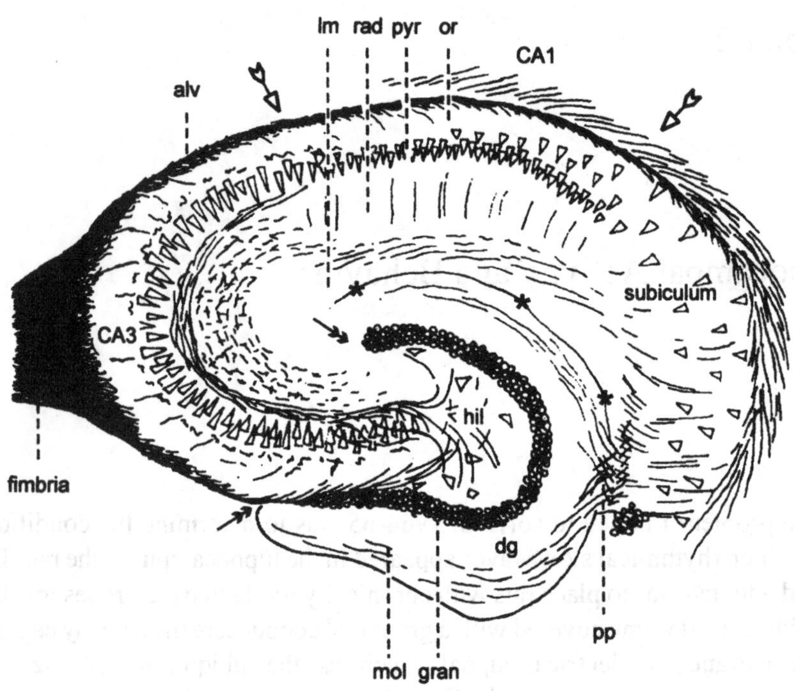

Figure 2-1. Drawing of a horizontal section through the hippocampal region in a rat. Modified from a paper published by T. Blackstad[1]. Solid double arrows mark the limits of the dentate gyrus; asterisks mark the position of the hippocampal fissure; hollow arrows mark the limits of field Cornu Ammonis I (CAI) or Ammon's horn I. Abbreviations: alv, alveus, a sheet of fibres covering part of the outer surface of Ammon's horn; CAI, CA3, subfields of Ammon's horn; dg, dentate gyrus; gran, granule cell layer of the dentate gyrus; hil, hilus of the dentate gyrus; lm, straum lacunosum-molecular of Ammon's horn; mol, molecular layer of the dentate gyrus (fine neuropil); or, stratum oriens of Ammon's horn; pp, perforant path, a major input to the dentate gyrus and Ammon's horn; pyr, stratum pyramidale of Ammon's horn; rad, stratum radiatum of Ammon's horn (a region containing the large apical dendrites of Ammon's pyramids). The fimbria is a major fibre pathway connecting the hippocampal formation with subcortical structures.

(e.g., in the skull over the cerebellum) the region near the fissure will be positive (and vice versa). If one records the difference in the electrical potential of these two sites, signals of up to 3 mV can be obtained. Since the two electrodes are less than 1 mm apart, the signals from other structures, such as the neocortex, are present at both of them and are not detected by the preamplifier (The common mode rejection ratio of modern differential preamplifiers is 10,000/1 or more). Consequently, a surface-to-depth bipolar electrode of this type yields a clear large amplitude hippocampal signal with few artifacts and minimal contamination from activity in other brain structures. The main facts are well illustrated in the elegant recordings taken by B. H. Bland and I.Q. Whishaw, which were published a few years later (see Figure 2-2).

When, at last, adequate slow wave signals from the hippocampus were recorded, their relation to behavior became very obvious. Gross movements such as walking, struggling to escape from my hand, or rearing up on the hind legs were invariably accompanied by rhythmical potentials of about 8-9 Hz but a more irregular pattern punctuated at irregular intervals by large spike-like potentials (sharp waves), occurred whenever the rat stood still. However, it also became apparent that a number of smaller movements such as turning the head, changing posture while resting, or moving a forepaw in isolation, were also reliably accompanied by rhythmical waves but both the amplitude and frequency (6-7 Hz) of these waves was less than it was during walking or struggling (Figure 2-3).

I spent many hours recording repeatedly from the same rats. Eventually, when they had become sufficiently accustomed to the situation, they would retreat to the back of the Faraday cage, lie down in a curled up posture and go to sleep. The neocortex, which usually gave rise to a low voltage mixed frequency record, (also known as low voltage fast activity) developed large amplitude, irregular slow waves. If a sudden stimulus was presented at such a time (a scratch on the screen of the cage or a hiss or whistle) the rat would leap to its feet, startled, its head up, eyes wide open, then stand motionless. This behavioral reaction was invariably associated with an abrupt replacement of the large slow waves of the neocortex with a low voltage higher frequency record (neocortical activation) but rhythmical waves did not appear in the hippocampus. Instead, the hippocampus assumed a pattern of irregular waves of very low amplitude. This indicated to me that the rhythmical hippocampal waves had nothing to do with arousal or alerting; they were specifically related to a class of movements that did not include the startle response. Furthermore, in the waking rat, sensory stimuli, in general, elicited hippocampal rhythmical slow activity only if they also elicited a certain type of motor activity. Ordinarily, a great variety of visual, auditory, tactile, and olfactory stimuli would elicit both hippocampal rhythmical

Figure 2-2. Distribution of hippocampal RSA in a hippocampal profile obtained during muscular paralysis (*d*-tubocurarine, 0.8 mg/kg) with artificial respiration and with pressure points and incisions treated with a local anesthetic (Xylocaine). Insert in upper left corner is a photomicrograph of a sagittal section of the hippocampal formation approximately 2.5 mm from the midline. The reference electrode is located in stratum moleculare, indicated diagrammatically by the open circle. The penetration path of the roving electrode is indicated by the long vertical line. Records of the activity at each numbered location correspond to the numbered records to the right of the section. (1) stratum oriens; (2) null zone in stratum radiatum; (3) stratum moleculare; (4) fast activity in hilus; (5) stratum moleculare lower blade. The record below the section with the open circle next to it represents the activity recorded from the reference electrode in stratum moleculare. Solid bars indicate the duration of the electrical stimulation (0.3 mA) of the posterior hypothalamus. Phase is shown by the Lissajous patterns taken from X Y plots. Left: RSA from point 1 in stratum oriens is approximately 180^0 out of phase compared with the reference in stratum moleculare. Right: RSA from point 3 in stratum moleculare is in phase with RSA recorded from the reference electrode in stratum moleculare. From Whishaw, I.Q., Bland, B.H., Robinson, T.E., and Vanderwolf, C.H. (1976). Neuromuscular blockade: the effects on two hippocampal RSA (theta) systems and neocortical desynchronization. *Brain Research Bulletin, 1*: 573-581. Reprinted with permission from Elsevier Science.

REF. ROVING

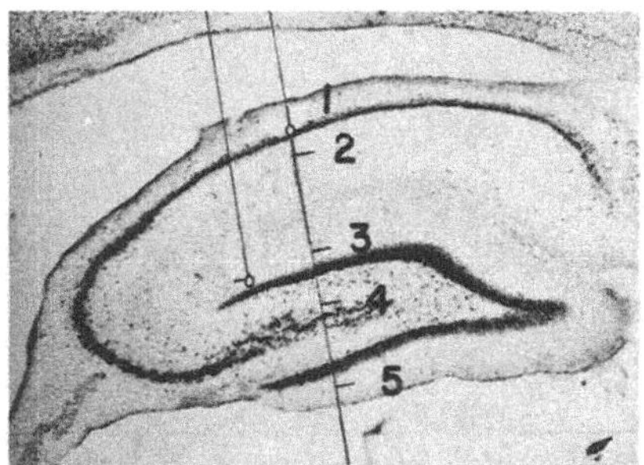

REFERENCE GENERATOR SITE

ST. MOLECULARE

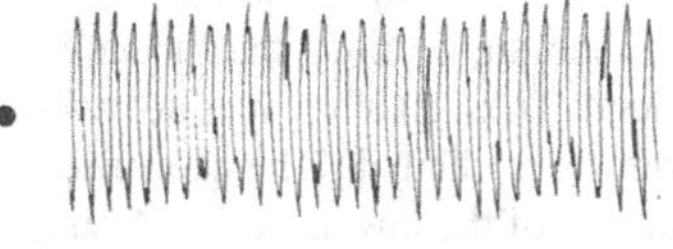

hp STIM. (.2 mA)

PHASE FROM X-Y PLOTS

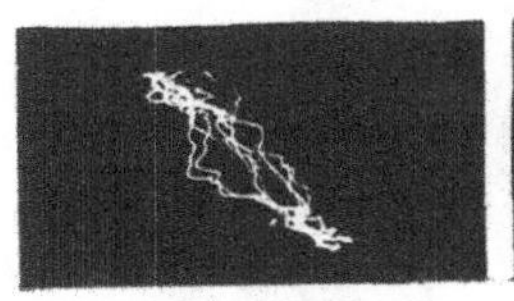

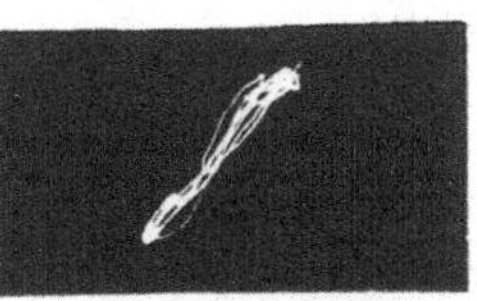

I vs REFERENCE 3 vs REFERENCE

ROVING CAI - DENTATE TRACK

•1

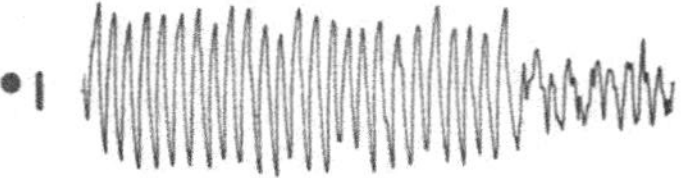

2

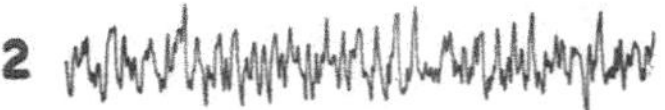

3

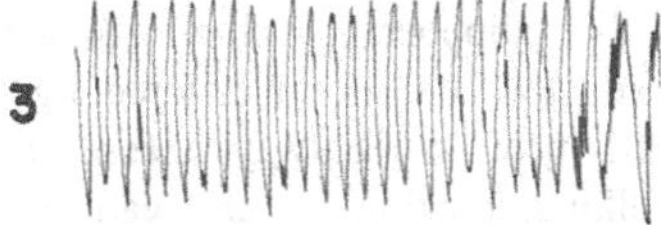

4

5

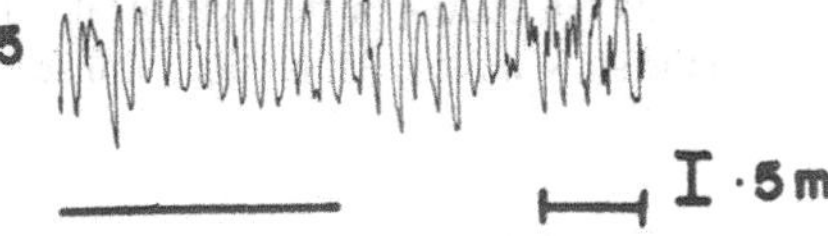

hp STIM. (.2 mA) I Sec.

Figure 2-3. Hippocampal slow wave activity in relation to behavior in rats. The traces shown were extracted from several different figures in a paper published in 1969 (Note #4). Rhythmical slow activity of 7-9 Hz always accompanies head movement occurring in an otherwise immobile rat, walking, picking up and carrying food in the mouth, struggling while held in the experimenter's hand, manipulating a food pellet while eating, and changing posture during an episode of grooming the fur, but more irregular activity occurs during waking immobility, quiet sleep, chewing food, washing the face with the forepaws and nibbling or biting the fur on the forelimbs. If a sleeping rat is suddenly awakened by a loud sound (Trace E) but it does not walk away or move its head about, a low amplitude irregular waveform is elicited. The neocortical trace in F is dominated by a spiky artifact associated with activity of the jaw muscles. A shift from biting the right foreleg to biting the left foreleg is associated with a burst of rhythmical slow activity but a pause in the rhythmical biting occurs with little or no change in the irregular pattern of the slow waves. Calibration: 1.0 second; 0.1 mV.

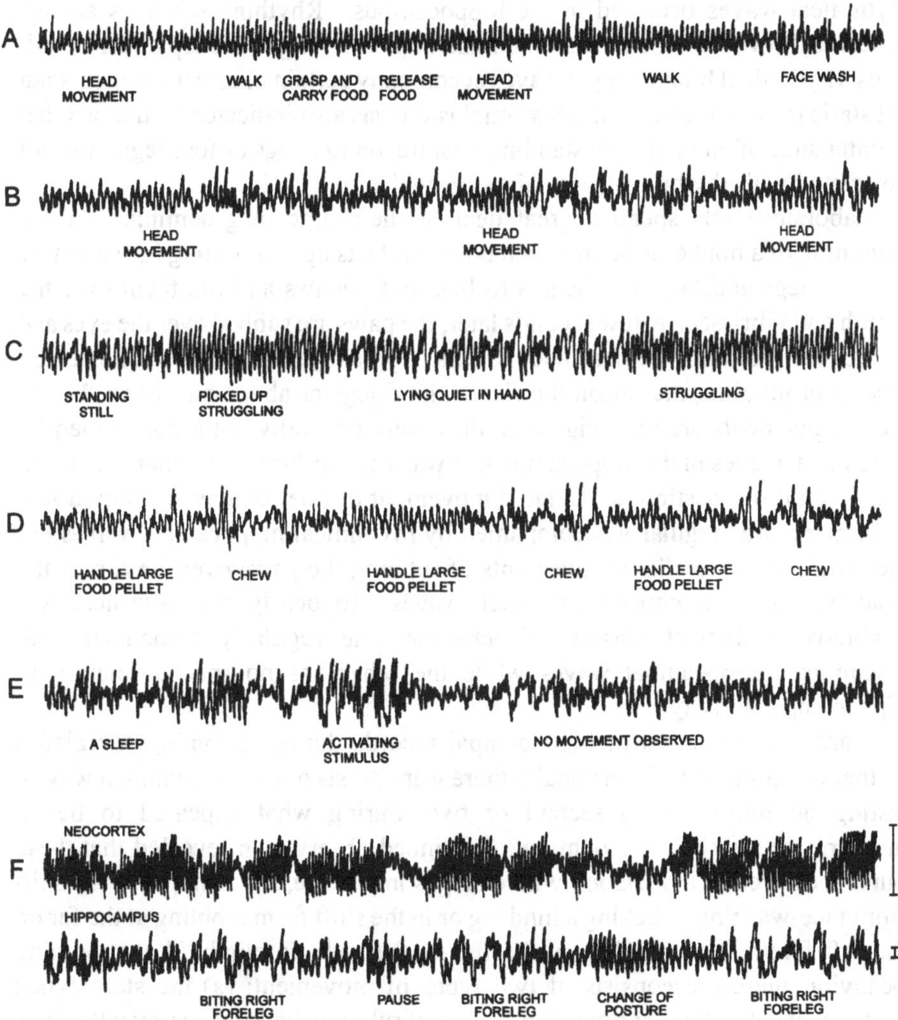
A
HEAD MOVEMENT
WALK
GRASP AND CARRY FOOD
RELEASE FOOD
HEAD MOVEMENT
WALK
FACE WASH
B
HEAD MOVEMENT
HEAD MOVEMENT
HEAD MOVEMENT
C
STANDING STILL
PICKED UP STRUGGLING
LYING QUIET IN HAND
STRUGGLING
D
HANDLE LARGE FOOD PELLET
CHEW
HANDLE LARGE FOOD PELLET
CHEW
HANDLE LARGE FOOD PELLET
CHEW
E
A SLEEP
ACTIVATING STIMULUS
NO MOVEMENT OBSERVED
F
NEOCORTEX
HIPPOCAMPUS
BITING RIGHT FORELEG
PAUSE
BITING RIGHT FORELEG
CHANGE OF POSTURE
BITING RIGHT FORELEG

slow activity and a behavioral response that included head movements, stepping, and locomotion.

Another instructive observation was made when the rats were placed on the vertical surface of a board clamped to the table. If a rat hung motionless, its weight supported mainly by its forepaws grasping the top of the board, no rhythmical waves occurred in the hippocampus. Rhythmical waves always appeared, however, when a rat pulled itself up, climbing to the top of the board. Thus, rhythmical hippocampal waves accompany certain phasic movements but not static muscular exertion. This conclusion was also indicated by the fact that maintenance of an immobile standing posture, on two legs or four legs, was not associated with rhythmical slow hippocampal activity.

Laboratory rats spend a great deal of their time in grooming. At the beginning of a bout of grooming behavior, a rat sits up, supporting its weight by the hind legs and tail, then begins to lick its forepaws and rub them over the mouth and vibrissae. A few seconds later, the paws are rubbed over the eyes and then over the back of the head and the ears. Typically, this phase is followed by licking or nibbling the fur on the flanks, hind legs or abdomen. Even though these movements are very vigorous, they were generally not accompanied by rhythmical waves in the hippocampus. It was very striking to see that when a rat was lying down resting, an isolated movement of one forepaw by only a few millimetres was regularly accompanied by rhythmical hippocampal waves but the much larger forelimb movements of rubbing the paws over the top of the head were not accompanied by such waves. Evidently, we have here two qualitatively distinct classes of behavior: one regularly associated with rhythmical hippocampal waves while the other has no special relation to hippocampal activity.

Careful observation of hippocampal records during grooming revealed a further puzzling fact. Occasionally there were bursts of clear rhythmical waves, lasting no more than a second or two, during what appeared to be an uninterrupted period of grooming. Continued observation revealed that these bursts occurred in association with changes in posture, for example in the shift from face-washing to licking a hind leg or in the shift from nibbling at the fur on the left forelimb to nibbling at the fur on the right forelimb. Rat grooming behavior, therefore consists of two types of movement: (a) the stereotyped movements of licking or biting the fur and of rubbing the forepaws over the face; and (b) changes in posture, a group of movements that can take a variety of forms depending on the posture at the beginning of the movement and the posture at the end of the movement. Only the latter type of movement is associated with rhythmical slow activity in the hippocampus.

All of these experiments were done with the experimenter (me) sitting very near the rat, in plain view, but quite motionless except for the movements of pressing the keys on the behavior recorder. Most mammals have very little reaction to a totally motionless human: if one sits immobile in the woods for a long time, mice, chipmunks, squirrels, or even deer may approach almost near enough to touch but a slight movement sends them galloping away. One day, while recording in this way, I shifted my weight in my chair momentarily as a rat was washing its face. The rat became totally immobile for several seconds, then continued face-washing. Curious, I stopped the paper drive and discovered that the irregular hippocampal activity characteristic of face-washing was not at all altered during the change in behavior. I repeated the observation many times with that rat and with others. The result, which was always the same, suggested to me that what is ordinarily referred to as perception and attention has little to do with the rhythmical waves of the hippocampus. It seemed that a rat could detect a movement (or other stimuli which I also tried) and react to it by becoming completely motionless (freezing behavior) without activating the hippocampus. However, if the rat's reaction to a stimulus included head movements or locomotion, rhythmical waves always appeared in the hippocampus.

I offered the rats food in the Faraday cage after depriving them of it overnight. Typically, a rat would stay well back in the cage, as far from the door as possible, but when a food pellet (roughly 5 g, 2-3 cm in length) was placed in the doorway, it would come forward slowly, sniffing all the while, and stretching out its body as much as possible, belly close to the floor, to permit reaching the food without coming any nearer the open doorway (and me) than was absolutely necessary. Then, in a sudden quick movement, the food was snatched up in the mouth and the rat would run to the back of the cage, sit up in a hunched posture, transfer the food pellet to the forepaws, bite off a piece, and begin to chew. When the first mouthful had been chewed and swallowed, the rat rotated the food pellet into a favorable position, took another bite and chewed again. This usually continued until the entire pellet was eaten, leaving only a little pile of crumbs on the floor.

Rhythmical waves occurred continuously in the hippocampus during the approach to food, during the snatch and run movements, and during the movement of manipulating the food with the forepaws, but hippocampal activity became quite irregular during the movements of chewing, provided that the rat remained otherwise motionless. When the pellet was large at the beginning of a bout of eating, the manipulatory movements were extensive, involving movement at the elbow and shoulder joints as well as the wrist and digits, but as the pellet became smaller, down to the size of a pea, the manipulatory movements were restricted largely to movements of the wrist and digits. Corresponding to this, the rhythmical waves in the hippocampus during manipulation were initially as well developed as they were during a large head movement but as the manipulatory movements were reduced in extent, the correlated hippocampal waves declined in amplitude and frequency.

When the rats were provided with water in a dish, I observed that the approach to the dish and turning away after drinking were always accompanied by rhythmical hippocampal waves but licking water was accompanied by irregular hippocampal activity.

The experiments on feeding permitted another observation relevant to a question that suggested itself. Rats often sniff as they move their heads and walk about, making rhythmical movements of the vibrissae as they do this. Therefore, it seemed possible that the rhythmical hippocampal waves were related to sniffing rather than to walking or head movement. I spent a considerable time watching for episodes of spontaneous walking unaccompanied by sniffing and satisfied myself that clear rhythmical slow hippocampal activity occurred at such times. However, it proved to be more difficult to find clear occurrences of sniffing occurring in isolation without accompanying head movement, walking or rearing up on the hind legs. The best approximation to this that I observed occurred when a rat had finished eating a large food pellet, leaving nothing but a tiny pile of crumbs on the floor under the snout. At this point, typically, a rat would lower its head and sniff at these crumbs very vigorously but make only very slight head movements. Little or no rhythmical wave activity occurred in the hippocampus during this behavior. Therefore, I concluded that the rhythmical hippocampal waves were related to head movements, locomotion, and the like, but had no particular relation to sniffing or vibrissae movement. It was also clear that the rhythmical hippocampal waves had no specific relation to exploratory behavior in a general sense. Handling food while eating or changing posture while grooming are very well practiced behaviors in a rat. They cannot be regarded as exploratory in any sense but they are, nonetheless, reliably associated with rhythmical slow hippocampal activity.

A waveform that was definitely related to sniffing could be recorded at sites in the dentate gyrus, especially the hilus (see Figures 2-1 and 2-2). High amplitude bursts of fast waves up to 50 Hz or so occurred very commonly during sniffing, a fact that impressed itself on me because of the distinctive rattling noise the ink-writing pens of the polygraph made at such times. I did not attempt to follow up this observation with any type of systematic investigation until about 25 years later (see Chapter 10).

The experiment that Woody Heron and I had done together in the summer of 1962 had shown that rhythmical waves could occur in the thalamus and hippocampus slightly in advance of overt motor activity. Observations of spontaneous behavior of the kind I had been making were not appropriate for

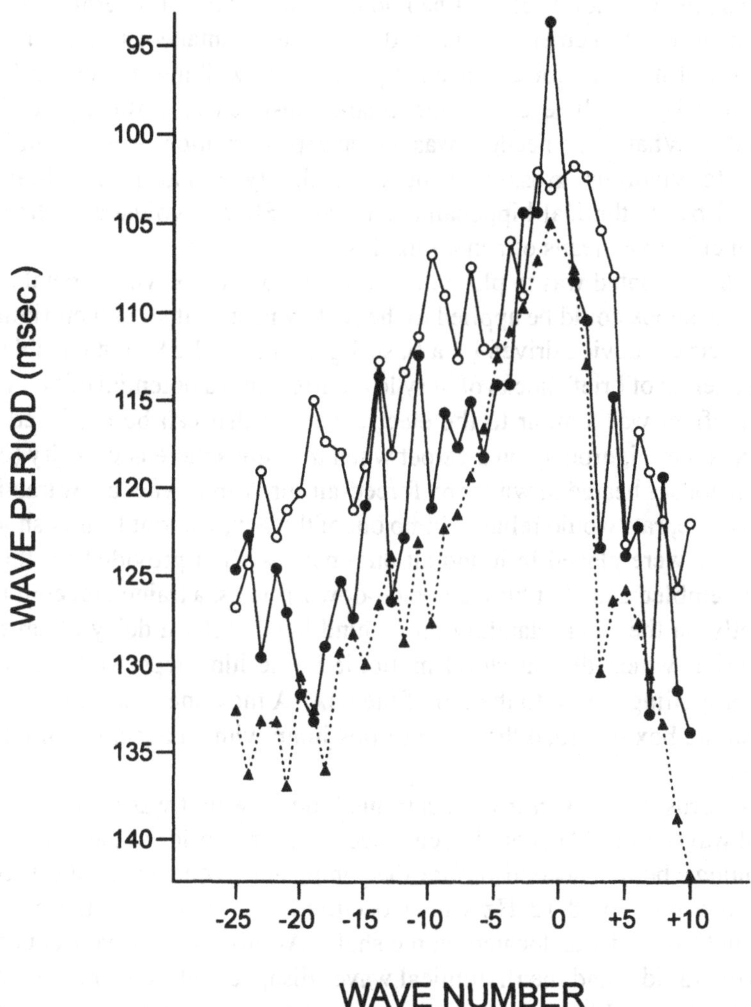

Figure 2-4. Changes in the frequency of hippocampal rhythmical slow activity associated with the behaviour of jumping out of a box. The jump begins at "0". The points shown are the means of 20-25 waves in each of three rats. A wave period of 100 milliseconds (msec) corresponds to a frequency of 10 Hz; a wave period of 135 milliseconds corresponds to a frequency of about 7.4 Hz. The frequencies shown are somewhat higher than is usual because the rats were warm from the exercise of repeated jumping. The frequency of hippocampal rhythmical slow waves increases by about 0.4 Hz / ^{0}C over the range of 26-42 ^{0}C (rectal temperature). From Vanderwolf, C.H. (1969). Hippocampal electrical activity and voluntary movement in the rat. *Electroencephalography and Clinical Neurophysiology, 26*: 407-418, with permission from Elsevier Science.

studying this phenomenon because I had no accurate means of determining the precise instant of movement onset and because spontaneous movements generally do not have an abrupt onset. Spontaneous walking, for example is usually preceded by small head movements and adjustments in posture (intention movements). What was needed was an abrupt transition from complete immobility to vigorous gross movement of the type that is consistently accompanied by rhythmical hippocampal waves. Shock avoidance behavior provided an effective means of achieving this.

The method adopted was to place the rat in a wooden box with a metal grid floor. Electric shock could be applied to the feet by means of an inductorium, a primitive electrical device driven by a flash-light battery (1.5V) but capable of producing a series of brief shocks of very low current and a potential of 500V or more. The effect was similar to the electric shocks that can be produced by rubbing one's feet vigorously on a carpet when the atmosphere is very dry (e.g. in Canadian houses heated in winter by forced-air furnaces). After a few training trials of this type, rats would reliably jump out of the box, without further shock, whenever they were placed in it and sit on a narrow shelf provided for them. (The box resembled a top hat turned upside-down.) Thus, a trained rat could be placed gently on the floor, standing on its hind legs. After a delay of several seconds during which the rat stood motionless, the hind legs would extend suddenly, propelling the rat to the top of the box. A movement-sensing device mounted on the box recorded the onset of this jump with an accuracy of a few milliseconds.

When records were taken from rats trained in this way it was observed that rhythmical waves of 6-7 Hz could occur several seconds prior to the jump, and that beginning about 1 second before the jump, the frequency of the waves increased to a peak of 8-12 Hz which occurred just at jump initiation and continued until the rat was located on the shelf. At this point, wave frequency declined very rapidly and the rhythmical waves disappeared as the rat resumed an immobile posture (Figure 2-4). It was an unvarying rule on every trial that if the rise in hippocampal rhythmical wave frequency failed to occur, the jump also failed to occur. If the rise in hippocampal wave frequency did occur, the rat would usually jump but sometimes it would walk to another part of the box instead. Thus, examination of hippocampal activity made it possible to predict the onset of movement with absolute certainty but not to predict what type of movement it was to be (walking or jumping).

The data from this experiment suggested that the hippocampus might have some role in both planning and the performance of a motor pattern. It also suggested a problem which subsequently became a major focus of my research. If the rhythmical waves of the hippocampus are related to motor activity, how is it possible that these waves can be present during relatively long intervals (several seconds) when a rat is absolutely motionless? The importance of this problem was further emphasized by research being conducted in an adjacent lab by Ron Harper[2], one of Woody Heron's graduate students. Ron had observed, and demonstrated to me, that in the hippocampus of New Zealand white rabbits, rhythmical waves could easily be elicited by various visual or auditory stimuli that produced no visible motor activity whatsoever. In normal rats this did not ordinarily occur. Stimuli that elicited rhythmical waves in the rat hippocampus also elicited motor activity, except in the case of the avoidance test situation.

Paradoxical or rapid eye movement (REM) sleep also presented an apparent exception to the general relation between hippocampal rhythmical slow activity and certain types of motor activity. The onset of REM sleep in a rat is always associated with an utter collapse of any pre-existing muscle tone. Thus, if a rat falls asleep in a crouched sitting posture, as they sometimes do, the onset of REM sleep is associated with the body slumping down limply on the floor. Despite this, bursts of muscular twitches occur periodically in the limbs, trunk, and especially in the vibrissae. Rhythmical slow waves occurred in the hippocampus throughout an episode of REM sleep, with higher frequency waves occurring during the muscular twitches than during the inter-twitch intervals. An interpretation of this curious phenomenon was suggested by research originating with Ottavo Pompeiano of the University of Pisa. It appears that brain motor systems generally are in a state of high activity during REM sleep but that overt expression of this activity is blocked by a powerful inhibition of spinal motor neurones and of reflex afferents to those neurons. Consequently, instead of running, jumping, etc. the animal lies limply on the floor, twitching slightly. The hippocampal record then, is related to motor activity during REM sleep as well as during waking.

This was approximately where the hippocampal research project stood in 1968 when I moved to the University of Western Ontario. My NRC grant had increased sufficiently to pay a post-doctoral stipend to Bob Sainsbury who had just completed a Ph.D. at McMaster. I had also accepted two talented graduate students from the Rod Cooper's lab in the University of Calgary, Brian Bland and Ian Whishaw. The four of us began an ambitious program of research with several related aims: (1) to extend the study of the correlations between spontaneous behavior and hippocampal slow wave activity; (2) to study the relation between hippocampal activity and behavior elicited by electrical stimulation of the hypothalamus and midbrain; and (3) to study the effects of a number of neuroactive drugs on hippocampal and neocortical activity in relation to behavior. The reasons for choosing these topics and the relations between them will, I hope, become apparent in the following chapters.

Scientific Reaction. As a young investigator hoping to achieve great things in brain-behavior research, I was truly entranced by all these observations. There was a great deal of discussion in the 1960s, as is still the case, of the possibility of relating brain activity to behavior. Well, I thought, here is a whole set of brain-behavior relations that can be demonstrated easily and clearly with very minimal equipment. I anticipated a very favorable response.

My first attempt to communicate these findings to the scientific world was to present a paper to a meeting of the Canadian Physiological Society[3]. My presentation was greeted with a long silence followed by one or two inconsequential questions which appeared to be related more to a desire to adhere to accepted social conventions than to any wish to elicit further information. The best question as I recall was: "How did you implant your electrodes?"

After the publication of an account of the experiments[4], negative opinions seemed to be intensified. As one senior investigator (Karl Pribram) put it at a later meeting (Tokyo, 1972): "Why would you ever think that the hippocampus has anything to do with motor activity?" Some of the commentators were less polite. When Ian Whishaw presented the main results of his Ph.D. thesis research at a meeting in the U.S., there were only a few inconsequential questions but after he had returned to his seat a woman sitting directly in front of him turned around and said, "That is the craziest stuff I've ever heard in my life"[5]. Under the cloak of anonymity, a referee of a paper submitted for publication wrote: "This is a good illustration of how not to do experiments," but did not offer any specific criticisms. Another referee, commenting on an application I had made for a research grant, wrote: "There is a great deal of this type of research done all over the world but its contribution to the progress of science is zero." Needless to say, my application was not successful. Such opinions made it difficult, for many years, to get papers published or to get research grants. However, I am very grateful that the National Research Council and later its daughter organization, the Natural Sciences and Engineering Research Council, continued to support my research at a level that permitted me to run a reasonably productive laboratory.

More encouraging results came from nearby investigators. Although I had several graduate students at McMaster (three of whom completed a Ph.D.) none of them worked on the hippocampal slow wave project because I judged that I did not then know enough about the field to provide adequate supervision. However, a colleague Abraham (Abe) H. Black had a graduate student named Art Dalton who was engaged in studies of conditioning in dogs. Art and Abe became very interested in the hippocampal slow wave work. It provided Art with a Ph.D. thesis[6] and several publications while Abe went on to carry out very elegant analytical studies of the relation between hippocampal activity and conditioned reactions[7]. In some of these experiments dogs and cats were trained to press a pedal with the forepaw (to avoid a mild electric shock) whenever a specific stimulus (a tone or a light) was presented and to refrain from pressing the pedal when a different stimulus was presented. Pedal pressing in the latter condition was punished by electric shock. Thus, under both conditions one might say that the animals were attentive to the stimuli, discriminated the stimuli, were motivated to avoid the shock, and had to remember what the stimulus meant. The two conditions seem to differ only in terms of the motor activity required: pedal pressing in the one case; immobility in the other case. Recording from the hippocampus, quantified by a computer-generated power spectral analysis or by an automated procedure of counting the occurrence of individual waves, showed that rhythmical slow waves were strongly associated with pedal pressing but rarely occurred during the immobility condition. These experiments, therefore, provided strong support for the idea that the rhythmical slow wave activity of the hippocampus is related in some way to the control of certain types of movement and not to supposed psychological processes such as attention, motivation, or memory.

Further, Abe Black and Art Dalton showed that large doses of tubocurarine which blocked neuromuscular junctions (completely paralyzing the animals and necessitating artificial respiration) did not prevent a well-established conditioned stimulus from eliciting hippocampal rhythmical slow activity. Therefore, this wave form cannot be due to proprioceptive feedback from motor activity. When naïve tubocurarine-paralyzed dogs were trained to produce rhythmical hippocampal waves (something that was easily done, it turned out) and the animals were tested again after recovery from the drug, it was found that they had acquired conditioned motor responses (lifting the head, stepping, etc). Therefore it is likely that hippocampal rhythmical slow activity in dogs under neuromuscular blockade is associated with strong activity in motor neurons but this is not seen, of course, because the muscles cannot respond.

Intellectual support from another direction was provided by the work of James B. Ranck Jr., then at the University of Michigan in Ann Arbor. Jim, recording single unit activity in the hippocampus of the freely moving rat, discovered a population of cells which fired at a high rate if, and only if, rhythmical slow waves were present in the hippocampus. Lower rates of firing occurred during the presence of irregular hippocampal slow waves. Therefore, the activity of these "theta cells" and the "theta rhythm"[8] were correlated with behavior in the same way. As Jim put it: "I have checked the firing of theta cells and slow waves against all of the observations in his (Vanderwolf's) 1969 paper, and have confirmed all his findings."[9] Needless to say, this type of confirmation was exceedingly welcome to me as an antidote to the negative or hostile criticism which my findings often aroused.

Notes on Chapter 2

1. Blackstad, T.W. (1956). Commisural connections of the hippocampal region in the rat, with special reference to their mode of termination. *Journal of comparative Neurology, 105*: 417-537.
2. Harper, R. (1968.) *Behavioral and electrophysiological studies of sleep and animal hypnosis*. Unpublished Ph.D. thesis, McMaster University, Hamilton, Ontario.
   Harper, R. M. (1971) Frequency changes in hippocampal electrical activity during movement and tonic immobility. *Physiology and Behavior, 7*: 55-58.
3. Vanderwolf, C. H. (1967). Behavioral correlates of "theta" waves. *Proceedings of the Canadian Federation of Biological Societies, 10*: 41-42.
4. Vanderwolf, C.H.(1969). Hippocampal electrical activity and voluntary movement in the rat. *Electroencephalography and Clinical Neurophysiology, 26*: 407-418.
5. Brian Bland and I (we were both present) each remember this story in this way but according to Ian, a woman stood up after his talk and said: "You mean to tell me that these waves always accompany swimming, running in a treadmill, lever pressing, and jumping out of boxes? I don't believe you!". Perhaps both of these stories are true, possibly referring to events occurring at different meetings. We all gave a lot of talks at meetings in those days and, after 30 years, one's recollections become a bit hazy. The specific words attributed to various people in this book are, at best, only approximations.
6. Dalton, A. (1968). *Hippocampal electrical activity in operant conditioning*. Unpublished Ph.D. thesis, McMaster University, Hamilton, Ontario.
7. Black, A. H. (1975) Hippocampal electrical activity and behavior. In R.L. Isaacson and K.H. Pribram (eds) *The hippocampus, volume 2: Neurophysiology and behavior*, New York: Plenum Press, 129-167.
8. The rhythmical slow activity of the hippocampus is often referred to informally as the "theta rhythm" or the "hippocampal theta rhythm". I had at first adopted these terms but gave them up when a journal editor insisted that the term "theta rhythm" was already in use to refer to a 4-7 Hz waveform occurring in human electroencephalography. Since the rat hippocampal rhythm can rise to a frequency of about 12 Hz, new terminology was required. I chose rhythmical slow activity (RSA).

9. Ranck, J. B. Jr. (1973). Studies on single neurons in dorsal hippocampal formation and septum in unrestrained rats. *Experimental Neurology*, *41*: 461-531.

Chapter 3

Hippocampal Slow Waves, Learning, and Instinctive Behavior

The idea that there is a localized cerebral area devoted to memory, and further, that the hippocampus is just such an area, had an enormous appeal to many people interested in brain-mind relations. Only two years after the appearance of Scoville and Milner's 1957 paper (see Chapter I) claiming that hippocampal lesions produced amnesia, E. Grastyan of the University of Pecs, in Hungary, published a study of hippocampal slow wave activity during learning in freely moving cats[1]. Grastyan and his colleagues reported that hippocampal rhythmical slow waves were characteristic of the early stages of learning when the cats displayed prominent orienting reactions (the what-is-it reflexes of Pavlov) but that both the orienting responses and the associated rhythmical slow waves disappeared when the learned behavior had become well established. Soon afterwards, W.R. Adey of the University of California at Los Angeles began publishing the results of a series of studies making use of computers to analyze changes in hippocampal slow waves during learning.

Although many people accepted these observations as a confirmation of the direct role of the hippocampus in learning and memory, it seemed to me that a different interpretation was more probable. Early in training, extensive exploratory movements such as walking, rearing, and head movement are likely to occur but later when the learned behavior is well established, unnecessary motor activity tends to disappear. I had observed that movements of small extent, such as turning the head or manipulating objects with the forelimbs were associated with rhythmical hippocampal waves of a lower frequency and smaller amplitude than the hippocampal waves associated with extensive movements such as walking or rearing. Therefore, the changes in hippocampal activity observed during training by Grastyan and Adey seemed to me to be related to the fact that the animals were doing different things before and after training. If one were to compare the electromyographic activity of somatic muscles during spontaneous behaviour before and after training, extensive changes would, no doubt, be observed but this would not mean that muscles are directly involved in learning and memory. Similarly, training-induced changes in hippocampal activity many be a consequence of a role played by the hippocampus in the control of motor activity.

The idea that hippocampal rhythmical slow waves would disappear during the performance of a well-practiced act seemed unlikely since such waves occurred prominently during the handling of food or the postural changes during grooming (Figure 2-3), actions which the rats had performed many thousands of times. Nonetheless, it seemed worthwhile to collect additional data. Ian Whishaw trained rats, cats, and Mongolian gerbils to run in motor driven treadmills or running wheels of appropriate size[2]. After preliminary training, he showed that hippocampal rhythmical slow activity was always present during continuous walking or running for periods of as long as 8-9 hr/day for several days. If the animals stopped momentarily, allowing themselves to be carried backwards on the moving treadmill, the hippocampal rhythmical waves were always interrupted. It was evident that long-continued practice does not necessarily result in a disappearance, or indeed any change at all, in the rhythmical slow waves of the hippocampus. When the same rats were trained to press a lever to obtain a small pellet of food, they walked about and reared a great deal early in training but soon settled down to a consistent pattern of lever pressing alternating with eating. Pressing the lever was reliably associated with rhythmical slow activity at good electrode sites (bipolar electrodes straddling the CAI pyramidal cell layer) but its frequency and amplitude were considerably less than the frequency and amplitude of the rhythmical waves associated with walking, rearing or running. A vertical jump to a height of 22 inches (~56 cm) was associated with rhythmical slow activity of a greater amplitude and a higher frequency than a jump of only 11 inches (~28 cm) (Figure 3-1). At recording sites placed more deeply in the hippocampus, the rhythmical waves associated with gross locomotor activity were quite obvious but the smaller amplitude waves associated with lever pressing could not be detected.

Ian and I concluded that these results provided a simple explanation of the findings of Grastyan and Adey[2]. Training is indeed capable of changing the frequency and amplitude of hippocampal slow waves and at some recording sites it may appear that these waves have disappeared altogether. These changes are related to the differing types of motor activity occurring at different times and have nothing directly to do with the training procedure; similar variation in the frequency and amplitude of hippocampal rhythmical slow waves accompany varying untrained spontaneous motor acts, as we have already seen.

In order to demonstrate that learning has an effect on hippocampal activity independent of changes in motor activity it is necessary to hold behavior constant during different training conditions. For example, one could compare hippocampal activity during spontaneous locomotion and during similar locomotion occurring as a result of training[3]. A number of comparisons of this type were published by us and by other research groups. If rats were trained to walk or run to find food or avoid shock, rhythmical slow activity resembling the rhythmical slow activity accompanying spontaneous walking and running was always present. If rats were trained to remain immobile, hippocampal activity was usually irregular, as it is during spontaneous alert immobility, but in some conditions a low frequency form of rhythmical slow activity occurred. Such activity usually had a frequency of about 6 Hz while the rhythmical slow activity accompanying certain movements ordinarily had a frequency of 7-9 Hz or more. This phenomenon was, naturally, of great interest to us and was investigated extensively (see Chapter 4).

During this period, W.R. (Ross) Adey and I were both speakers at a conference on the hippocampus organized by the Neurosciences Research Program at Woods Hole in Massachusetts in 1976. Naturally, we did not agree on the question of the behavioral correlates of hippocampal rhythmical slow activity. After a prolonged period of what is known in the trade as a "vigorous discussion", it was agreed that a special evening session should be held to try to resolve the dispute.

The high point of this discussion was a presentation by Per Andersen of the University of Oslo who was easily the most eminent hippocampal electrophysiologist of the time. Per rose to describe a film made by Fernando Lopez de Silva of the Institute of Medical Physics in Utrecht. Hippocampal slow wave activity had been recorded in a dog trained in a sort of traffic-light experiment in which one light signalled the dog to walk around a circular track while a second light signalled stop. Per, playing the role of the dog, walked in a circle in a crouched posture while describing the hippocampal pattern verbally. "So" said Per, walking, "theta-theta-theta-theta-theta, stop! No theta." As he said this he stopped and stood erect, holding his hand up, palm outward, in the manner of a policeman directing traffic. Then crouching again and walking, he continued, "Theta-theta-theta-theta-theta-theta, stop! No theta."

Shortly afterwards, the special evening meeting ended. Jim Ranck, another of the participants in the conference, told me afterwards, laughing, that Per's

Figure 3-1. Hippocampal slow wave activity in relation to various behaviors in a rat. This Figure is comparable to Figure 2-3 but additional behaviors are represented and the quality of the record is better. Note: (a) rhythmical slow activity during active or paradoxical sleep, with an increase in both frequency and amplitude when movements occur; (b) higher amplitude and frequency when jumping 22 inches (56 cm) than when jumping 11 inches (28 cm); (c) rhythmical slow activity during handling and during swimming; (d) irregular activity during waking immobility (sitting still) chattering the teeth and during quiet sleep; (e) rhythmical slow activity with a reduced amplitude and frequency during a small movement of the head; and (f) suppressed irregular activity when a sleeping rat is awakened when the experimenter (Ian Whishaw) taps a pencil on the table. From Whishaw, I.Q., and Vanderwolf, C.H. (1973). Hippocampal EEG and behaviour: changes in amplitude and frequency of RSA (theta rhythm) associated with spontaneous and learned movement patterns in rats and cats. *Behavioral Biology*, *8*: 461-484.

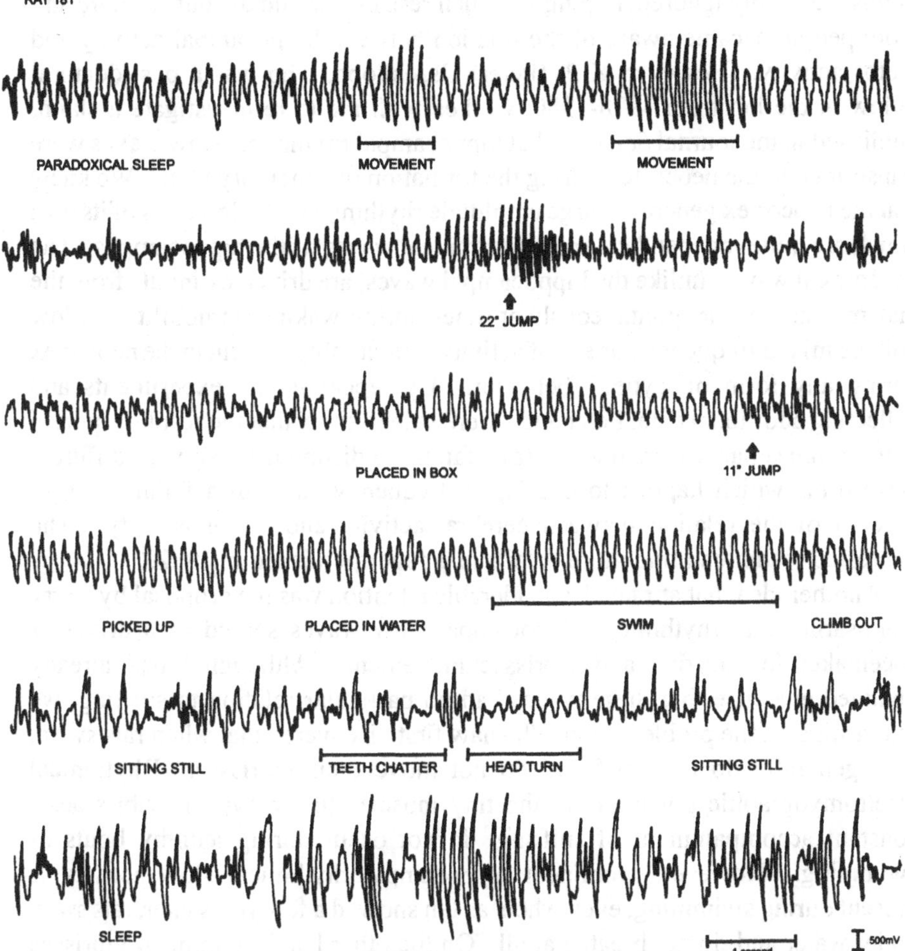

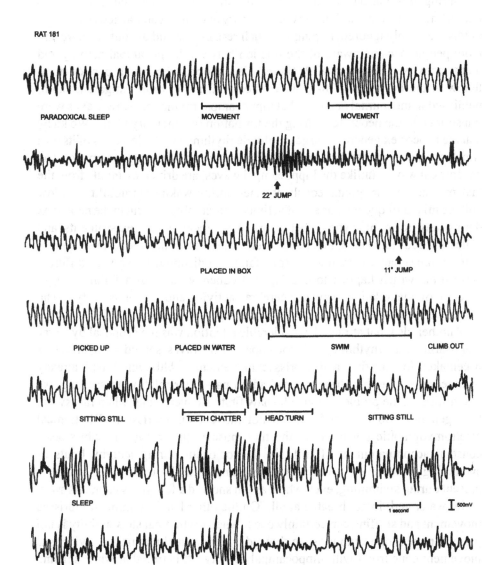

RAT 181
PARADOXICAL SLEEP
MOVEMENT
MOVEMENT
22" JUMP
PLACED IN BOX
11" JUMP
PICKED UP
PLACED IN WATER
SWIM
CLIMB OUT
SITTING STILL
TEETH CHATTER
HEAD TURN
SITTING STILL
SLEEP
1 second
500mV
SLEEP
OBSERVER TAPS PENCIL

demonstration was the most impressive evidence yet produced that hippocampal theta waves were related to motor activity.

During this period, various other investigators carried out experiments purporting to show a relation between memory and hippocampal activity. Much of this we simply ignored, hoping that such research would die out as more and more people became aware of the relation between hippocampal activity and motor activity. However, K.P. (Peter) Ossenkopp, who was a post-doctoral fellow in the lab from 1978-1980, worked with me to re-investigate a claim, published in the journal *Science*, that hippocampal rhythmical slow waves were transmitted to the neocortex during the formation of a memory trace. We knew that the neocortex generates large amplitude rhythmical 7-9 Hz waves of its own which are quite independent of the rhythmical waves of the hippocampus. The neocortical waves, unlike the hippocampal waves, are driven by inputs from the thalamus and occur spontaneously at times during waking immobility. A low voltage mixed frequency pattern of activity is invariably present in the neocortex during any form of Type 1 behavior. By repeating the experiments and collecting additional data, Peter and I were able to show that the results reported in the *Science* paper were due to: (a) a failure to distinguish two very different waveforms which happen to overlap in frequency; and (b) a failure to take account of the relation between cerebral activity and motor activity. The interested reader can find more details in a paper published in 1982.[4]

Another idea that attracted considerable attention was the proposal by Barry Komisaruk[5] that rhythmical hippocampal slow waves served as a driver or pacemaker for sniffing and vibrissae movement. Although I had already obtained evidence that these waves had no particular relation to sniffing, we reinvestigated the problem. Ian Whishaw first discovered that when rats swim, they generally do not sniff and do not move their vibrissae. Rhythmical electromyographic potentials in the tiny muscles that move the vibrissae, a constant accompaniment of sniffing, do not occur during lengthy bouts of swimming. However, prominent hippocampal rhythmical slow activity is present during swimming, even when, as Ian showed a few years later, rats swim under water and do not breathe at all. On the other hand rhythmical vibrissae movement and sniffing can certainly occur when rhythmical slow activity is not present in the hippocampus as we showed by recording breathing and vibrissae movement concurrent with hippocampal activity. It is true that when sniffing and hippocampal rhythmical slow activity occur concurrently there is often a relation between the two rhythms, a fact that appears to be attributable to the relation between rhythmical slow activity and the head movements that usually accompany sniffing. The interested reader can find a more extended discussion of this issue in a paper published in 1992[6].

An important conclusion suggested by the correlations of behavior with spontaneous hippocampal activity is that there is little difference between "learned' and "instinctive" behavior. This point was emphasized by Ian Whishaw's work and by Bob Sainsbury's studies of hippocampal activity during agonistic behavior, courtship and mating in guinea pigs[7]. Hippocampal rhythmical slow activity accompanied walking or running under all conditions regardless of whether it occurred in a guinea pig alone in a box, in a male guinea pig engaged in an aggressive encounter with another male, or in a male guinea pig pursuing a female. In contrast, hippocampal activity was predominantly irregular, not only during alert immobility but also during such behaviors as ano-genital licking or pelvic thrusting during copulation.

Over 20 years later, Larissa Mead, a graduate student prompted, perhaps, by the spirit of the times, pointed out that the earlier studies by Sainsbury and others had not given due attention to hippocampo-behavioral relations in females. Larissa showed[8] that female rats display high levels of hippocampal rhythmical slow activity during hopping (part of the estrus display of female rats) but more irregular activity during lordosis , a posture in which a female rat stands immobile, her rump tipped upward and the tail deviated to one side to permit vaginal penetration by the male. During parturition, irregular hippocampal activity was also present during the stretching and whole body writhing movements accompanying birth, as well as during the licking of the new born pups or the mother's own genitalia. The writhing movements of parturition, like the startle response, are examples of large scale powerful movements that are not consistently associated with hippocampal rhythmical slow activity. Rhythmical slow activity was always present during the behavior of carrying the pups or on any other occasion when locomotion occurred.

One of the most remarkable facts in the entire brain-behavior field is that brief repetitive pulses of electric current delivered via a wire electrode thrust into one or another of the various regions of the hypothalamus can elicit most of the spontaneous behaviors observed in the species under study. Stimulation at some sites will cause a rat to seek out a block of wood and gnaw large splinters from it while an adjacent piece of food is ignored. Stimulation at other sites may elicit eating, drinking, predatory attacks, grooming, copulation, or digging holes in earth or sawdust. In contrast, stimulation of the motor cortex, for example, tends to elicit unnatural looking movements such as rotation of the head accompanied by a repetitive flailing movement of the contralateral forelimb. This suggested a question: would the behaviors elicited by hypothalamic stimulation be accompanied by patterns of hippocampal activity similar to those that accompany the spontaneous occurrence of the same behaviors?

Brian Bland and Ian Whishaw both carried out experiments in this field while they were graduate students[9]. Brian solved the difficulty of recording from a running rat tethered to the ends of short recording and stimulating leads by placing the rat in a light plastic running wheel that rotated on a ball bearing. Electrical stimulation of the posterior hypothalamus produced vigorous running or galloping accompanied by large amplitude rhythmical slow activity. It was possible to demonstrate quantitative relations between the strength of the stimulating current, the frequency of the elicited rhythmical slow activity and the speed at which the rat ran. These findings suggested a possible dependence of the elicited behavior on ascending projections to the hippocampus, a topic to which Brian and his students returned over two decades later. Oddie and Bland[10] demonstrated that injection of small quantities of procaine (a local anesthetic) in the medial septal region abolishes the elicitation of both hippocampal rhythmical slow activity and wheel running behavior by electrical stimulation of the posterior hypothalamus. This suggests that both effects are dependent on an ascending hypothalamo-septo-hippocampal pathway. Stimulation of the lateral hypothalamus produced somewhat different effects. Low-level stimulation at sites that produced eating or drinking (in rats that were satiated and would not eat or drink spontaneously) produced clear rhythmical slow activity in the hippocampus as the rat walked toward a food pellet or toward a water spout, but irregular activity occurred as the rat chewed or licked. Rhythmical slow activity was always present during stimulation-elicited digging (in sawdust) just as it is during spontaneous digging. We concluded that the behavior patterns elicited by hypothalamic stimulation are accompanied by the same sort of hippocampal activity as are spontaneous behaviors of the same kind.

Interesting effects of an entirely different type were observed when the hippocampus itself was stimulated electrically. Repetitive stimulation of Ammon's horn usually produced seizure patterns: high amplitude sharp waves and fast waves in the hippocampus accompanied by periods of immobility, convulsive movement without a loss of the normal standing posture, and wet-dog shakes. This was succeeded by a period in which hippocampal activity was largely suppressed accompanied by locomotor hyperactivity. In contrast, stimulation of the dentate gyrus rarely gave rise to epileptic phenomena when stimulus trains were confined to durations of not more than a few seconds. Brian Bland[11] discovered that stimulation of the dentate hilus in the middle of the septo-temporal hippocampal formation produced short latency evoked potentials over an extensive septotemporal region of CAI. This suggests activation of a widely projecting monosynaptic (or oligosynaptic) system. Repetitive stimulation at low frequencies had very little behavioral effect, but at higher frequencies (5-100 Hz), when each spontaneous rhythmical slow wave cycle was interrupted by one or more evoked potentials (thereby disrupting normal function), there was a complete arrest of spontaneous locomotion, lever pressing, swimming, jumping out of a box (a previously established avoidance response), or struggling in response to being held. These effects were obtained in the absence of any electrographic or behavioral signs of seizure activity. The motor activities of maintaining a normal standing posture, licking movements of the tongue (drinking water) and shivering (after immersion in cold water) appeared to be quite unaffected by dentate stimulation. Therefore, as a first approximation, dentate gyrus stimulation selectively disrupts Type 1 behaviour but has little effect on Type 2 behaviour.

These were important observations. The fact that hippocampal rhythmical slow waves occur in close correlation with certain patterns of movement does not necessarily mean that hippocampal activity has a role in causing the movement. It is well recognised that correlation does not prove causation. However, the dentate stimulation experiment provided evidence that a disruption of normal hippocampal activity produced a loss of the Type 1 behavior to which hippocampal activity was related. This finding strongly supported the idea that hippocampal activity is not merely correlated with behavior: it actually has a role in generating behavior.

It is apparent that some motor patterns, including the various forms of locomotion, head movements, spontaneous changes in posture, and manipulating objects with the forelimbs, are invariably accompanied by hippocampal rhythmical slow activity, while other motor patterns, including alert immobility, licking, biting, chewing, face-washing, and such gross motor patterns as the startle response and the writhing-stretching movements of giving birth, are generally accompanied by an irregular pattern of hippocampal activity. These hippocampo-behavior relations occur during both spontaneous behavior and the behavior elicited by hypothalamic stimulation. The two different classes of behavior cannot be distinguished on the basis of extent of muscular activity, degree of arousal, stress, or excitement and have no particular relation to the often stressed polarity of learning and instinct. How should all this be interpreted and what should these classes of behavior be called?

Animal behaviorists, following a proposal by Wallace Craig in 1918, often distinguish appetitive from consummatory behavior. Walking toward food is an appetitive behavior; eating the food, a consummatory behavior. Prior to Craig's suggestion, Charles Sherrington[12] had suggested a distinction between precurrent reactions (similar to Craig's appetitive behavior) and consummatory reactions, stressing the dependence of the first type on distance receptors (vision, audition, olfaction) and of the second type on contact receptors (touch, taste). However John Hughlings Jackson[13], an English neurologist writing well before either Sherrington or Craig, had suggested a continuum in the basis of motor control ranging from most voluntary to most automatic or reflexive. I had become acquainted with Jackson's ideas during my years at McGill. Penfield was a great admirer of Jackson: a bust of the neurologist stood in the main lecture theatre in the Montreal Neurological Institute. Consequently, I began to refer to behaviors consistently accompanied by hippocampal rhythmical slow activity as "voluntary" and the behaviors not consistently accompanied by this wave form as "automatic". However, the attempt to apply Jacksonian terminology to hippocampo-behavioral relations was not welcomed. "How can you call running in a treadmill voluntary?" asked one referee. Consequently, having little wish to debate terminology, I adopted "Type I behavior" as the designation for behaviors that are always accompanied by hippocampal rhythmical slow activity and "Type 2 behavior" for behaviors that have no consistent relation to hippocampal rhythmical slow activity. It is important to keep in mind that rhythmical slow activity may occur at times during Type 2 behavior. For example, rhythmical slow activity occurs if a rat takes a step forward during an uninterrupted period of licking water from a dish. If a female guinea pig takes a step or two forward during copulation, the male will display hippocampal rhythmical slow activity as he hops forward to maintain his position. The irregular hippocampal activity occurring during sniffing movements is converted to rhythmical slow activity when sniffing is associated with head movements or locomotion. A summary figure illustrating the main features of hippocampal slow wave activity in relation to behavior is shown in Figure 3-2.

Although philosophical discussions about behavior are fascinating, running an active laboratory demands a strongly practical approach. Brain-behavior research requires an enormous amount of hands-on work. In addition to the actual experiments, animals have to be cared for; recording and stimulating electrodes have to be manufactured and surgically implanted; brain sections have to be prepared and stained; supplies have to be ordered; accounts have to be kept, equipment has to be bought or built, repaired, and so forth. We had the good fortune to have an excellent animal technician in Bob Davis, a man whose main career was painting watercolors of the scenery of southern Ontario. This work, appealing though it was (and ultimately quite successful), could not pay the bills – hence animal care as a source of steady income. The need for someone to build and repair equipment was met by hiring John Orphan, a skilled machinist who later provided the foundation for an excellent group of technicians in the Department of Psychology.

Finding a suitable surgical and histological technician was another problem. Those I had tried fell into two groups: (1) those who performed well but soon moved on to other things, such as graduate school; and (2) those who, because of carelessness, unreliability or downright deceit in concealing mistakes (a truly unforgivable sin) were encouraged to find some other line of work. In 1970, I was once again looking for a technician. Shinshu Nakajima, an old friend who

HIPPOCAMPUS **BEHAVIOR**

Type 1

walking, running, swimming, rearing, jumping, digging, manipulation of objects with the forelimbs, isolated movements of the head or one limb, shifts of posture.

Related terms: voluntary, appetitive, instrumental, purposive, operant, or "theta" behavior.

Type 2

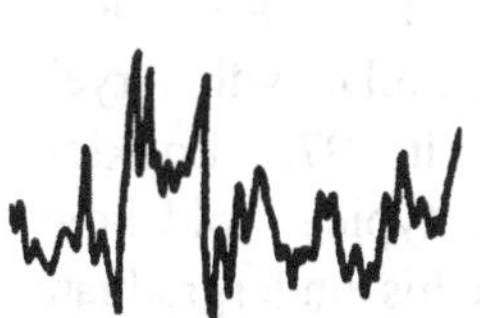

a) alert immobility in any posture.
b) licking, chewing, chattering the teeth, sneezing, startle response, vocalization, shivering, tremor, face-washing, scratching the fur, pelvic thrusting, ejaculation, defecation, urination, piloerection.

Related terms: automatic, reflexive, consummatory, respondent, or "non-theta" behavior.

Figure 3-2. Summary of hippocampal slow wave activity in relation to behavior in the rat. Type 1 behavior is always accompanied by rhythmical slow activity but Type 2 behavior is ordinarily accompanied by irregular wave activity. From Vanderwolf, C.H., Kramis, R., Gillespie, L.A., and Bland, B.H. (1975). Hippocampal rhythmic slow activity and neocortical low-voltage fast activity: relations to behavior. In R.L. Isaacson and K.H. Pribram (eds.) *The hippocampus, volume 2: Neurophysiology and behavior*, New York: Plenum Press, 101-128. With permission from Kluwer Academic/Plenum Publisher.

had been a fellow graduate student at McGill, recommended a candidate to me. On October 15, 1970, my office door opened to reveal a rather slight young man with reddish blond hair. He walked in very boldly, shook my hand and announced: "Hello! I'm Richard Cooley!" Thus began a long and productive association. Richard performed most of the surgery and histology done in the laboratory, became an excellent photographer and maker of illustrations for publications and made important improvements in technical matters such as the manufacture of recording leads and movement sensors. He also improved and centralized record keeping by introducing the "Great Book," a series of large format volumes that devoted one page to each rat that passed through the surgery, providing details on what was done to the rat and, after its death, what was done to its brain. Every October 15th, for 28 years, Richard would come to my office or to the lab to shake my hand saying: "Hello! I'm Richard Cooley!".

Bob Sainsbury moved on to a tenurable position at the University of Calgary in 1969. In 1971, Brian Bland and Ian Whishaw each successfully defended a Ph.D. thesis. Ian found himself a tenurable position at the newly founded (1967) University of Lethbridge in Alberta; Brian accepted a post-doctoral fellowship with Per Andersen at the University of Oslo in Norway. Their places in the hippocampal project were taken by Craig Milne and by Larry Gillespie (another graduate student from Rod Cooper's lab at the University of Calgary). Ron Kramis, a native of Montana who had just completed a Ph.D. with Aryeh Routtenberg at Northwestern University in Chicago, arrived in 1972. Looking back on these events, it is easy to see the operation of a typical "old boys network". I had known Aryeh at McGill where he took his undergraduate degree; Rod Cooper had been a fellow graduate student in Hebb's laboratory.

Notes on Chapter 3

1. Grastyán, E., Lissak, K., Madarasz, I., and Donhoffer, H. (1959). Hippocampal electrical activity during the development of conditioned reflexes. *Electroencephalography and clinical Neurophysiology 11*: 409-430.
2. Whishaw, I.Q., and Vanderwolf, C.H. (1973). Hippocampal EEG and behavior: changes in amplitude and frequency of RSA (theta rhythm) associated with spontaneous and learned movement patterns in rats and cats. *Behavioral Biology 8:* 461-484.
3. Jim Ranck once suggested that this procedure should be called "behavior clamping" since it is logically analogous to the voltage clamp procedure employed to study changes in current uncomplicated by variations in voltage during the occurrence of an action potential.
4. Vanderwolf, C.H., and Ossenkopp, K.-P. (1982). Are there patterns of brain slow wave activity which are specifically related to learning and

memory? In: C. Ajmone Marsan and H. Matthies (eds). *Neuronal plasticity and memory formation*, New York: Raven Press, pp. 25-35.

5. Komisaruk, B. (1970) Synchrony between limbic system theta activity and rhythmical behaviors in rats. *Journal of Comparative and Physiological Psychology 70*: 482-492.
6. Vanderwolf, C.H. (1992) Hippocampal activity, olfaction and sniffing: an olfactory input to the dentate gyrus. *Brain Research, 593*: 197-208.
7. Sainsbury, R.S. (1970). Hippocampal activity during natural behavior in the guinea pig. *Physiology and Behavior, 5*: 317-324.
8. Mead, L.A., and Vanderwolf, C.H. (1992). Hippocampal electrical activity in the female rat: the estrous cycle, copulation, parturition, and pup retrieval. *Behavioral Brain Research, 50*: 105-113.
9. Bland, B.H., and Vanderwolf, C.H. (1972). Diencephalic and hippocampal mechanisms of motor activity in the rat: Effects of posterior hypothalamic stimulation on behavior and hippocampal slow wave activity. *Brain Research, 43*: 67-88.
   Whishaw, I.Q., Bland, B.H., and Vanderwolf, C.H. (1972). Hippocampal activity, behavior, self-stimulation, and heart rate during electrical stimulation of the lateral hypothalamus. *Journal of Comparative and Physiological Psychology, 79*: 115-127.
10. Oddie, S.D., & Bland, B.H. (1998). Hippocampal formation theta activity and movement selection. *Neuroscience and Biobehavioral Reviews, 22*: 221-231.
11. Bland, B.H., and Vanderwolf, C.H. (1972). Electrical stimulation of the hippocampal formation: Behavioral and bioelectrical effects. *Brain Research, 43*: 89-106.
12. Sherrington, C. (1906). *The integrative action of the nervous system.* New Haven: Yale University Press.
13. Taylor, J. (1958). *Selected writings of John Hughlings Jackson, volumes 1 and 2.* London: Staples Press.

Chapter 4

Two Afferent Systems Control the Activation of the Neocortex and Hippocampus

Throughout most of the twentieth century, it was very widely believed that the pattern of slow wave potentials recorded from the neocortex (the electrocorticogram) or from the surface of the scalp (the electroencephalogram) is closely related to the level of consciousness. This concept is illustrated in Figure 4-1 taken from Penfield and Jasper's 1954 book on "*Epilepsy and the functional anatomy of the human brain*"[1]. In general, high levels of consciousness or excitement were said to be correlated with relatively low voltage higher frequency (fast) potentials while sleep or unconsciousness were said to be correlated with higher voltage lower frequency (slow) potentials. I was skeptical about this, in part because of doubts about the validity of psychological interpretations of *any* cerebral events, and in part because certain well-established facts did not agree with the conventional theory. One of these facts was the discovery by A. Wikler in 1952 that atropine, a drug that blocks some of the effects of the neurotransmitter acetylcholine, produces an abundance of large amplitude slow wave activity in the electrocorticogram in dogs without producing behavioral sleep or coma.

In a classic paper in 1914, Henry Dale, a British pharmacologist, showed that application of acetylcholine to various tissues has two broad classes of effects: (a) muscarinic effects which consist of excitatory or inhibitory actions on smooth muscle, the heart, and certain glands; and (b) nicotinic effects, which consist primarily of excitation of skeletal muscle. The term "muscarinic" was adopted

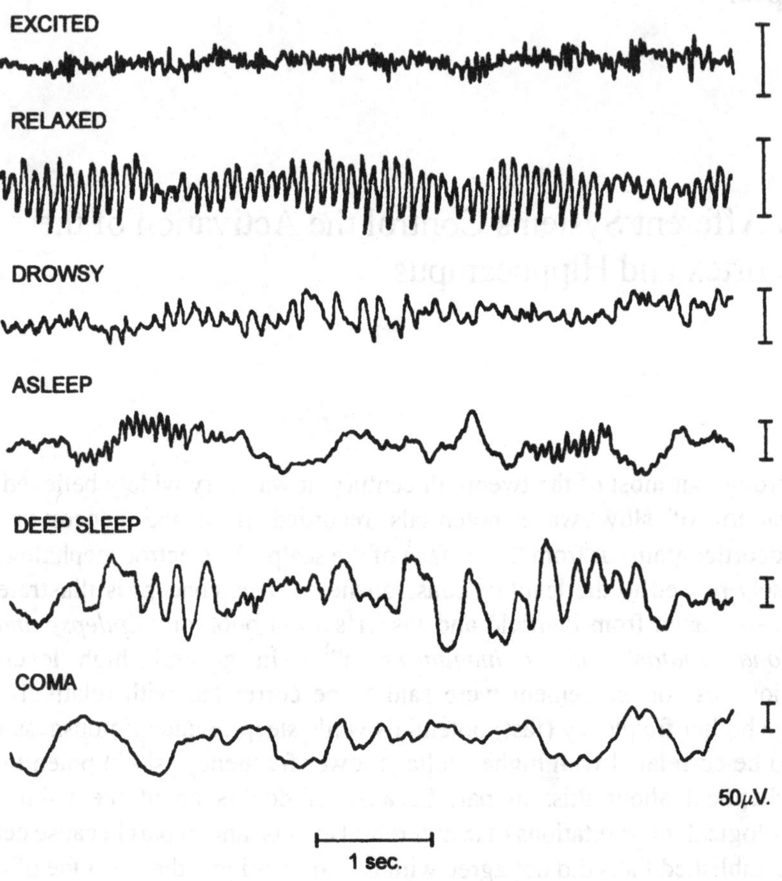

Figure 4-1. The relation between the electroencephalogram and the level of consciousness according to Penfield, W., and Jasper, H. (1954). *Epilepsy and the functional anatomy of the human brain*. Boston: Little, Brown & Co (reprinted with permission).

because effects of this type can be elicited by muscarine, a chemical present in *Amanita muscaria,* which is a poisonous mushroom formerly used to kill flies (musca in Latin). Similarly, the "nicotonic" effects of acetylcholine can be duplicated by nicotine, present in tobacco.

Atropine, the drug used by Wikler, is a classical antimuscarinic agent, an antagonist of the muscarinic effects of acetylcholine. In addition to its effects on autonomic function such as dilatation of the pupil, paralysis of accommodation, blockade of salivation and sweating, and increases in heart rate, atropine has pronounced behavioral effects such as the production of a delirious state which is afterwards not remembered (amnesia). Being quite intrigued by all this when I first learned about it in a course in pharmacology at McGill, I obtained some atropine sulfate and tested its effect on the acquisition of avoidance behavior in rats. The drug produced a severe impairment in acquisition but only in doses much larger than those required to block peripheral autonomic effects. This, as I learned later on, is largely due to the fact that atropine does not penetrate the brain nearly as well as it penetrates other tissues and also because rats possess a special enzyme (atropinesterase) that rapidly hydrolyzes atropine. However, it was (and is) reasonable to believe that the behavioral effects of atropine are largely due to a blockade of the muscarinic effects of acetylcholine in the brain.

When I recorded the electrocorticogram in rats before and after systemic injection with atropine sulfate, I confirmed Wikler's observations: atropine produces large amplitude slow waves, similar or identical to those occurring in natural sleep, in an animal which is awake and hyperactive. I also observed an effect not mentioned by Wikler or by others who had subsequently worked on this topic. The electrocorticogram in an atropinized rat is very strongly related to behavior[2]. Large slow waves occurred during periods of immobility but a low voltage fast record (also called neocortical activation), similar to what is seen during normal waking behavior, was present whenever the rat moved its head or walked about (Figure 4-2). As in the case of recording rhythmical slow activity in the hippocampus, the neocortical effect was visible in monopolar records (active electrode on the surface of the neocortex or penetrating to a depth of 0.5-1.0 mm; indifferent electrode over the cerebellum) but the clearest effects were observed in bipolar surface-to-depth records in which the deep electrode penetrated to a depth of nearly 1.0 mm (to approximately layer 5). This is due to the fact that the large slow waves of the neocortex show a phase reversal with depth: when surface waves are negative, waves recorded in layer 5 are usually positive and vice versa.

Research is often described as a problem-solving activity: one must design experiments to answer a clearly formulated question. However, it often happens that one observes a phenomenon which looks interesting because of an intuitive recognition that it provides the answer to an important question, but without any

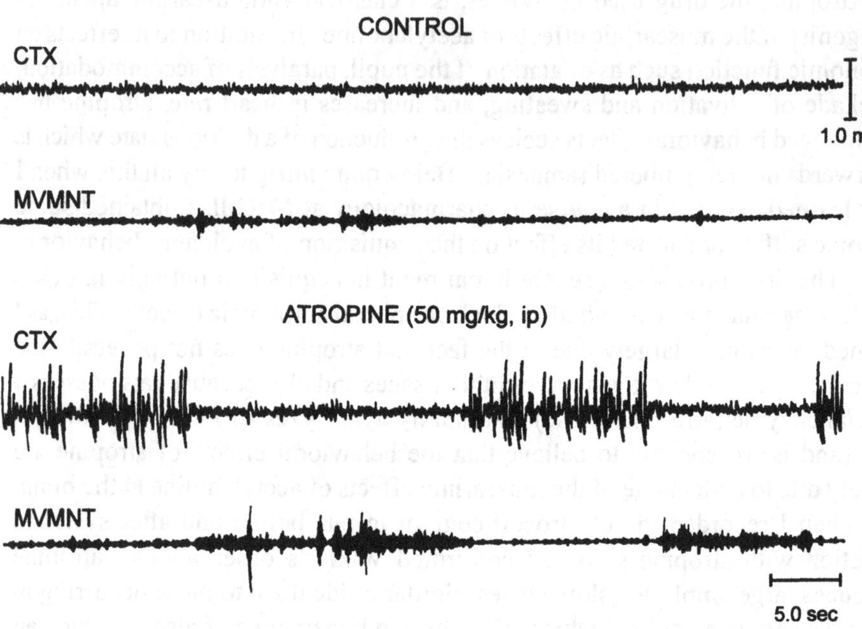

Figure 4-2. Effects of atropine sulfate on neocortical electical activity in relation to behavior. CTX, neocortical wave activity; MVMNT, output of magnet-and-coil type of movement sensor. *Upper traces*; undrugged rat displays continuous low voltage fast activity (LVFA) with no relation to motor activity. *Lower traces*; after atropine sulfate (i.p.) large amplitude slow waves occur during immobility but LVFA persists during spontaneous motor activity (head movements and stepping). From Vanderwolf, C.H. (1984). Aminergic control of the electrocorticogram: a progress report. In A.A. Boulton, G.B. Baker, W.G. Dewhurst and M. Sandler (eds.) *Neurobiology of the trace amines*. Clifton, New Jersey, 163-183. With permission from Humana Press.

explicit understanding of exactly what the question is. One must then discover the question. What does the phenomenon of behavior-related wave activity in the neocortex of an atropinized rat actually tell us?

I studied this phenomenon at length[3]. The large slow waves occurred not only during behavioral immobility but also during face-washing, spontaneous tremor, and chattering or gnashing of the teeth. Other behaviors such as chewing food or drinking water could not be studied because they are blocked by atropine. Nonetheless, it was apparent that the large slow waves occurred during Type 2 behavior while a low voltage fast record was present during Type 1 behavior. Therefore, some aspect of neocortical activation must be related to behavior in the same way that hippocampal activation (i.e. rhythmical slow wave activity) is related to behavior. Curiously, this was not apparent in neocortical records in undrugged rats. As Figure 4-2 shows, a waking rat usually displays a low voltage fast neocortical record both when it moves its head or walks about and when it stands motionless. One could say that an intact rat has an excellent ability to maintain a low voltage fast neocortical record during Type 2 behavior but that this ability is lost after treatment with atropine.

Further study revealed a number of other interesting facts about the atropine-induced behavior-related waves. First, the phenomenon occurred throughout the neocortex, not just in the sensori-motor areas as one might be inclined to suspect. It the electrodes were positioned correctly, the phenomena revealed in Figure 4-2 could also be demonstrated in the striate (visual) cortex and in the temporal (auditory) cortex. This indicates that some aspect of the activity of the neocortex, as well as the hippocampus, is organized in relation to Type I behavior. It may be that the occipital cortex functions as a visuo-motor area, the lateral or temporal cortex as an audio-motor area, the frontal and parietal cortex as a tacto-motor or proprio-motor area and that all these areas have a special function relevant to the control of Type I behavior. It is important to remember that the large pyramidal cells of layer 5, which provide the efferent outflow of the neocortex to all subcortical structures except the thalamus, are also the main sources of the electrocorticogram. These topics are discussed more extensively in Chapter 8.

The electrocortical effects produced by atropine are also produced by a variety of other drugs with a central anti-muscarinic action including scopolamine hydrobromide, quinuclidinyl benzilate, promethazine hydrochloride and Ditran but not by a variety of other drugs with different pharmacological actions. Atropine and scopolamine molecules that have a second methyl group attached to the nitrogen atom (atropine methyl nitrate, scopolamine methyl bromide) retain a peripheral anti-muscarinic effect but do not penetrate the blood-brain barrier. These drugs have little or no effect on the electrocorticogram or on behavior. These facts indicate that the electrographic and behavioral effects of atropine and scopolamine are due to an anti-muscarinic action in the brain.

As in the case of hippocampal rhythmical slow activity, the neocortical activating effect persisting after atropinization is related to the actual behavior of the rat and not to presumed causes of behavior such as stimulus input or arousal state. The following observations provide a simple demonstration of this. Atropinized rats are hyperactive, spending most of their time moving the head about, stepping, or walking. The electrocorticogram consists of continuous low voltage fast activity, interrupted by a burst of slow waves whenever the rats stop moving momentarily. A sudden strong stimulus, such as dropping a metal waste-paper basket on the floor, produces either: (a) a startle response followed by continued or even intensified motor activity; or (b) a startle response followed by a brief period of immobility (freezing behavior). In the case of outcome (a) there is no clear change in the ongoing pattern of neocortical low voltage fast activity but in the case of outcome (b) the stimulus-induced immobility is always accompanied by a burst of large slow waves that disappear only when the rat begins moving again. Stimulus-induced freezing behavior is usually regarded as a sign of high arousal or fear. If so, this phenomenon provides an instance in which a strongly arousing stimulus elicits large amplitude neocortical slow waves, demonstrating that atropine-resistant neocortical activation is related to overt behavior, not to "arousal", regarded as a state of internal excitement.

One obvious possibility suggested by all these observations is that the neocortex in an atropinized rat is activated (i.e., a low voltage fast pattern is elicited) by proprioceptive feedback from movement. One way of testing this idea is to compare active movements, initiated by the rat, with passive movements made by the experimenter. This is difficult to do because atropinized rats are highly reactive, moving about and often vocalizing whenever they are touched. However, if a major tranquilizer (an antipsychotic drug such as trifluoperazine or haloperidol) is administered, the hyperactivity and hyper-reactivity are abolished but the atropine-resistant activation of the neocortex continues to occur normally whenever the rat makes an active movement. Under these conditions it is possible to move the limbs, head or whole body of an atropinized rat without eliciting either an active movement or neocortical activation. Atropine resistant hippocampal rhythmical slow activity (see below) is not elicited by passive movement either. Some years later, Brian Bland, Terry Robinson, Ian Whishaw and I showed that these wave forms can also occur under conditions of neuromuscular blockade in which movement and proprioceptive feedback from movement cannot occur. This indicates that atropine-resistant neocortical and hippocampal activation is not dependent on feedback from movement. Quite possibly, the neural activity underlying these wave patterns plays a role in the causation of active movement.

In other experiments, hippocampal and neocortical activity were recorded concurrently in rats that had been trained to jump out of the avoidance apparatus, then giving them atropine[4]. The result was rather confusing to me. It was obvious that the amount of rhythmical slow activity displayed by the hippocampus during an avoidance response was very substantially reduced but it was not wholly abolished. Neocortical activation was also present during avoidance behavior in an atropinized rat. I did not feel much enlightened by this experiment.

Meanwhile, Ron Kramis had been working industriously on the behavioral correlates of hippocampal rhythmical slow activity in New Zealand White rabbits. Like rats, the rabbits always displayed the rhythmical waves during head movements, spontaneous changes in posture, and locomotion but, unlike rats, they also often displayed the same rhythmical slow waves during behavioral immobility or face-washing, especially in response to sensory stimuli such as whistling, slamming doors, flashing lights, or light touches to the skin. Ron and I made use of electromyographic recording from various neck, trunk, and limb muscles to convince ourselves that this immobility-related rhythmical slow activity could occur without any increase in muscular activity. Up to this time, I had thought that all rhythmical slow activity occurring during immobility had a low frequency (about 6 Hz) indicating a low level of activation of central motor systems and that greater activation with higher frequency rhythmical slow waves was required for movement to occur. This assumption was based on data obtained in the avoidance situation (Chapter II). However the rabbits soon showed that this was quite wrong. Rhythmical slow waves of up to 12 Hz could be obtained in response to sensory stimuli without the slightest indication of any Type 1 movement. Ron became very skeptical of the whole idea that hippocampal rhythmical slow waves had any particular relation to motor activity. Things looked very confused and discouraging.

In situations like this, the best thing to do is to make more observations even though one has no real understanding of what is going on. The summer of 1973 provided me with the opportunity to work in the laboratory from late April to early September. During the September to late April period of each year, teaching and the paper-shuffling associated with administrative duties (not to mention the time spent writing papers and grant applications) always absorbed an inordinate amount of time, precluding any serious research. The summer of 1973 turned out to be one of the most productive of my entire research career.

A phenomenon which impressed me very much at the time was that animals in surgical anesthesia with volatile anesthetics, such as diethyl ether, often display spontaneous neocortical activation and hippocampal rhythmical slow activity. This phenomenon, first described for the neocortex by F. Bremer in Brussells, Belgium, in 1936, seems to establish that the presence of activated

Figure 4-3a. Effect of lateral hypothalamic stimulation on behavior and on hippocampal and neocortical electrical activity. Stimulation changes irregular hippocampal activity into 8-10 Hz rhythmic slow waves and elicits head movements and walking. Neocortical activity maintains a low voltage fast pattern throughout. Under the conditions of these experiments, each volt of the applied stimulation provides a current of about 20 microamperes. Movement of the rat is detected by observation and by a magnet-and-coil movement detector. Calibration: 1s time intervals, 500 μV. Figures 4-3a-c are reprinted from: Vanderwolf, C.H., Kramis, R., Gillespie, L.A. and Bland, B.H. (1975). Hippocampal rhythmic slow activity and neocortical low-voltage fast activity: relations to behavior. In R.L. Isaacson and K.H. Pribram (eds.). *The hippocampus, volume 2: Neurophysiology and behavior*, New York: Plenum Press, 101-128, with the permission of Kluwer Academic/Plenum Publishers.

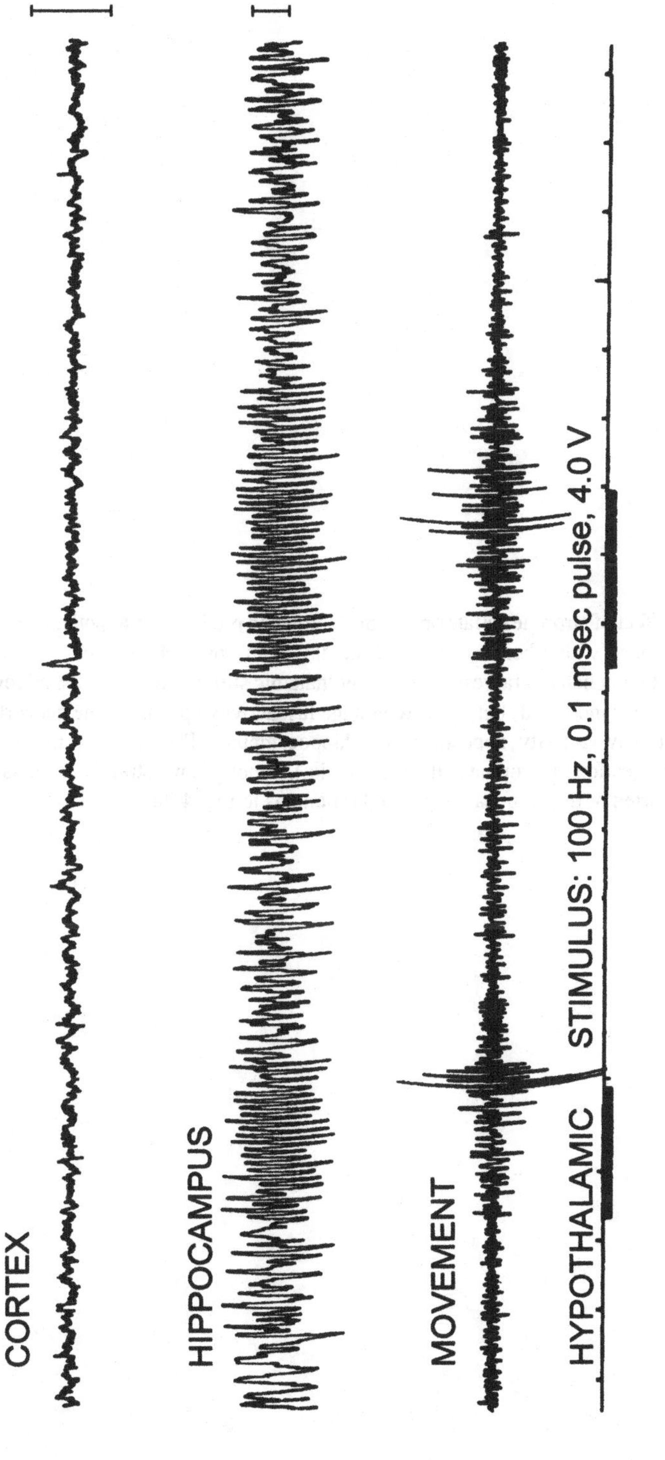
CORTEX
HIPPOCAMPUS
MOVEMENT
HYPOTHALAMIC STIMULUS: 100 Hz, 0.1 msec pulse, 4.0 V

Figure 4-3b. Effect of atropine sulfate on cerebral activation produced by hypothalamic stimulation in a rat anesthetized with ether. Same rat, same day as in Figure 4-3a. Left: Rat under ether anesthesia. Note that behavioral response to hypothalamic stimulation is abolished (even though stimulus intensity is increased) but that low-voltage fast activity appears in the neocortex and 5-6 Hz rhythmical slow activity appears in the hippocampus. Right: Ten minutes after an intraperitoneal injection of atropine sulfate (50 mg/kg), hypothalamic stimulation fails to activate either the neocortex or the hippocampus. Calibration: as in Fig. 4-3a.

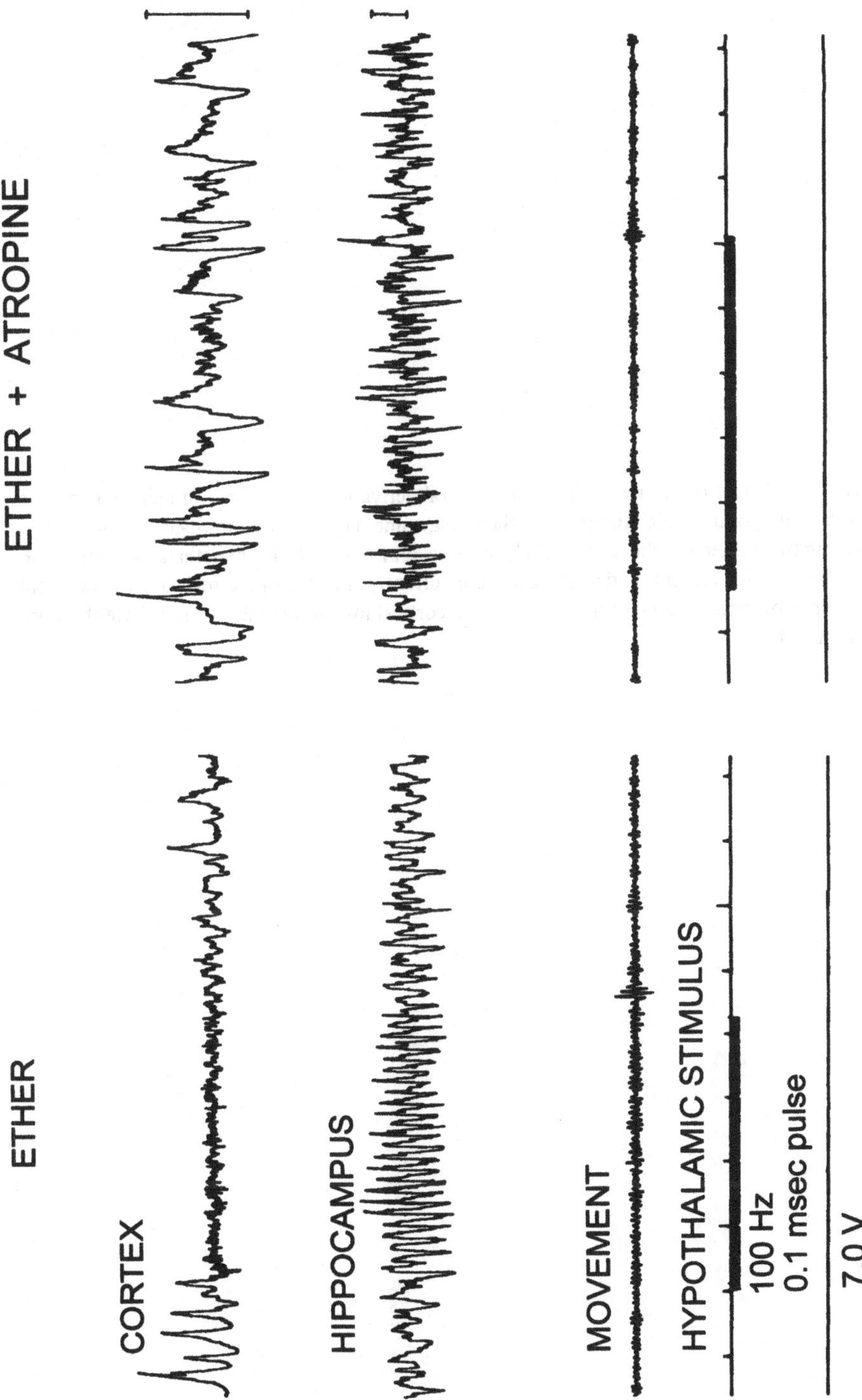
ETHER
ETHER + ATROPINE
CORTEX
HIPPOCAMPUS
MOVEMENT
HYPOTHALAMIC STIMULUS
100 Hz
0.1 msec pulse
7.0 V

Figure 4-3c. Effect of atropine sulfate on cerebral activation accompanying behavior in a rat during lateral hypothalamic stimulation. Same rat, same day as in Figures 4-3a and 4-3b. Rat recovering from ether anesthesia, but still heavily atropinized. Note that hypothalamic stimulation now elicits head movements and walking and concurrently elicits atropine-resistant hippocampal rhythmical slow activity and atropine-resistant necortical low-voltage fast activity. Calibration: as in Figure 4-3a.

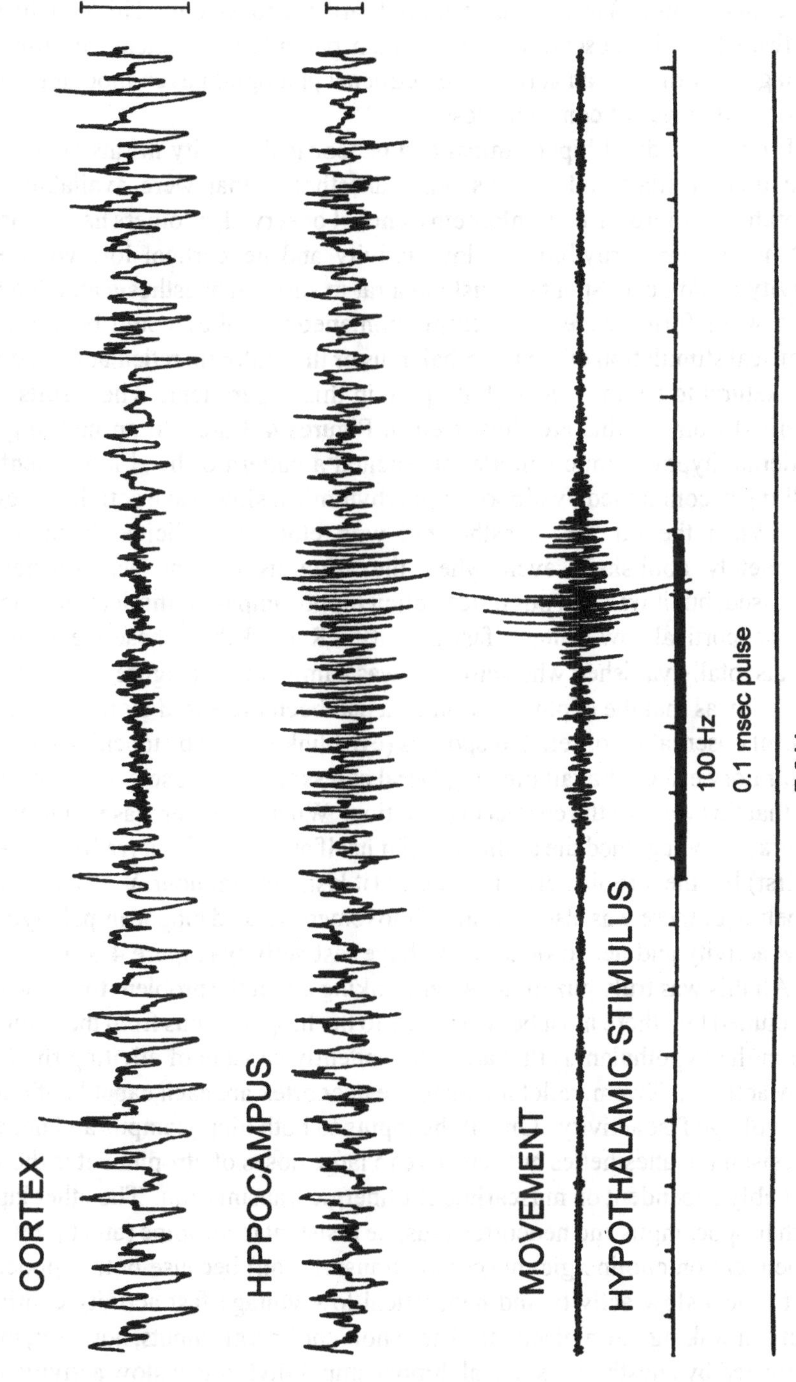
ATROPINE
CORTEX
HIPPOCAMPUS
MOVEMENT
HYPOTHALAMIC STIMULUS
100 Hz
0.1 msec pulse
7.0 V

cerebral patterns does not necessarily indicate the presence of consciousness since (one assumes) anesthetized animals are not conscious. This conclusion is confirmed by the observation that human patients with extensive brain stem damage may display an activated electroencephalogram even though they give no sign whatever of consciousness.[5]

I took records of hippocampal and neocortical activity in rats anesthetized with diethyl ether and various other anesthetics that were available in the laboratory (chloroform, trichloroethylene). I observed, as others had before me, that hippocampal rhythmical slow activity and neocortical low voltage fast activity could occur spontaneously in a rat in surgical anesthesia and that when these wave forms were not present spontaneously, they could be elicited by electrical stimulation of the hypothalamus or the midbrain reticular formation. It was natural to try the effect of atropine in this experiment. The results which emerged from all this are illustrated in Figures 4-3 a-c. In an undrugged rat moderate hypothalamic stimulation elicited a pattern of head movements and walking accompanied by hippocampal rhythmical slow wave activity (Figure 4-3a). When the rat was anesthetized with ether, the elicited behavior was completely abolished, even when the stimulus current was substantially increased, but it was still possible to elicit hippocampal rhythmical slow activity and neocortical low voltage fast activity. Remarkably, these electrographic effects totally vanished when atropine was administered (Figure 4-3b). My first thought was that the depth of anesthesia had been increased by the atropine but careful observation of reflex responses (eye blink elicited by touching the cornea, respiratory effect of a tail pinch, general assessment of muscle tone) convinced me that this was not the correct explanation. When the ether was withdrawn, the rat gradually regained the ability to right itself and walk about (rather unsteadily at first) but the atropine effect persisted (it lasts several hours). With the return of behavior there was also a return of movement-related hippocampal rhythmical slow activity and neocortical low voltage fast activity (Figure 4-3c).

All this was truly puzzling. After thinking about the problem for some time, I concluded that there must be two inputs to the hippocampus from the stimulation site in the hypothalamus that are independently capable of eliciting rhythmical slow activity. Two parallel inputs to the neocortex are each capable of eliciting low voltage fast activity. One of the inputs to both hippocampus and neocortex is resistant to anesthetics but sensitive to large doses of atropine: it is therefore, probably dependent on muscarinic cholinergic transmission. The other inputs to both hippocampus and neocortex must be resistant to atropine (and therefore not dependent on cholinergic muscarinic transmission) because both hippocampal rhythmical slow activity and neocortical low voltage fast activity continue to occur in waking atropinized rat. These non-cholinergic inputs, however, must be abolished by anesthetics since all hippocampal rhythmical slow activity and all neocortical low voltage fast activity disappears in an anesthetized rat given atropine.[6]

I already knew that the neocortical low voltage fast activity that was abolished by atropine occurred during Type 2 behaviors such as immobility and face-washing. Perhaps the hippocampal rhythmical slow activity that was sensitive to atropine was also the type that was present during immobility. Anesthetized rats are immobile, after all. Then it occurred to me that I had already made a test of this idea in unanesthetized rats. In the avoidance situation rats display hippocampal rhythmical slow activity both during jumping and during the period of immobility that precedes jumping. Hastily, I looked again at the records from the avoidance experiment. What I had not been able to see before was now obvious. Long trains of rhythmical slow activity did not precede jumping in the atropinized rats: the hippocampal wave form occurred only in close association with the jump itself[7]. Therefore, it might really be true that the hippocampus has an atropine-sensitive (presumably cholinergic) form of rhythmical slow activity which can occur, at times, during behavioral immobility and other Type 2 behavior and an atropine-resistant form of rhythmical slow activity which occurs only during the performance of Type 1 behavior such as head movement or walking.

Ron and I immediately tested this idea in his rabbits. We began by taking records of hippocampal activity during spontaneous hopping and during the presentation of various visual, auditory, and tactile stimuli in a motionless rabbit. Atropine sulfate was injected and the tests were repeated after a delay of a few minutes. The results were clear the first time we tried the experiment. Although it was easy to elicit rhythmical slow activity in the hippocampus during immobility prior to atropinization, such waves no longer appeared after atropine was injected (see Figure 4-4). In contrast, the rhythmical slow waves accompanying hopping were not altered in any obvious way by atropine. Rhythmical slow activity occurring during spontaneous face-washing behavior also proved to be sensitive to atropine. Consequently it appeared that in the rabbit, as in the rat, hippocampal rhythmical slow activity occurring during immobility or other Type 2 behavior is sensitive to central muscarinic blockade (atropine-sensitive) while the similar waves occurring during locomotion and other Type 1 behavior are only slightly affected by muscarinic blockade (atropine-resistant). A further curious phenomenon which we observed was that visual or auditory stimuli produced a slight suppression or other alteration in breathing prior to atropinization. This effect persisted after atropine was injected, indicating that the blockade of atropine-sensitive rhythmical slow activity was not due to a generalized depression of brain reactivity.[8]

Another interesting test of the difference between the rhythmical hippocampal waves occurring during immobility and those occurring during

Figure 4-4. Effects of atropine sulfate on hippocampal electrical activity in a rabbit during hopping and during sensory stimulation. Note that sensory stimulation (stroboscopic flashes, indicated by a heavy black line) produced changes in respiration both before and after atropinization, but did not elicit movement. Hippocampal rhythmic slow response to sensory stimulation is abolished by atropine (injected in the marginal vein of the ear) while the rhythmical slow pattern accompanying hopping persists unchanged. Calibration: 1 s, 500 μV. From Kramis, R., Vanderwolf, C.H., and Bland, B.H. (1975). Two types of hippocampal rhythmical slow activity in both the rabbit and rat: relations to behavior and effects of atropine, diethylether, urethane, and pentobarbital. *Experimental Neurology, 49*: 58-85.

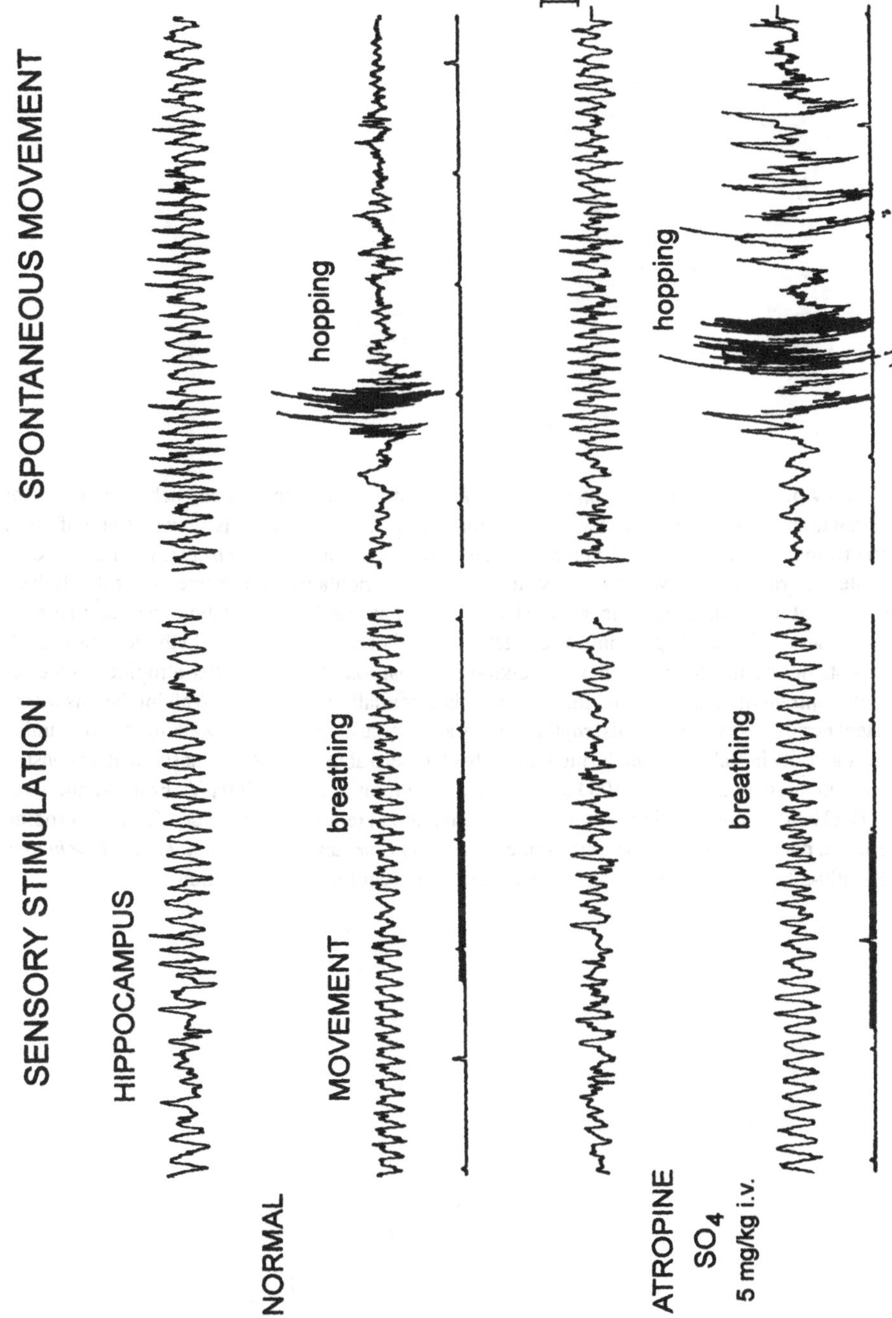

SENSORY STIMULATION
SPONTANEOUS MOVEMENT
HIPPOCAMPUS
NORMAL
MOVEMENT
breathing
hopping
ATROPINE
SO4
5 mg/kg i.v.
breathing
hopping

Figure 4-5. Effects of atropine on hippocampal rhythmical slow activity and behavior elicited by stimulation of the midbrain reticular formation. Hippocampal activity is shown as an unfiltered record and also after passage through a band pass filter system. *Top four tracings*: Undrugged rat. Note that rhythmic slow activity accompanies small spontaneous movements (probably head movements) and that rhythmic slow activity begins at onset of stimulation even though rat is motionless. Walking begins only after a latent period of about 3 s. Also, rhythmic slow activity persists during the centrally elicited walking. *Bottom four tracings*; After atropine sulfate (25 mg/kg intraperitoneally). Note that behavior is essentially unchanged. Walking begins after a latent period of several seconds; rhythmic slow activity still accompanies walking, but rhythmical slow activity initially present during immobility has been abolished. Reticular stimulation constant throughout (0.1 ms pulses at 100 Hz, and 4.5 V). Calibration: 1.0 s., 500 μV. From Vanderwolf, C.H. (1975). Neocortical and hippocampal activation in relation to behavior: effects of atropine, eserine, phenothiazines and amphetamine. *Journal of Comparative and Physiological Psychology*, *88*: 300-323 with the permission of the American Psychological Association.

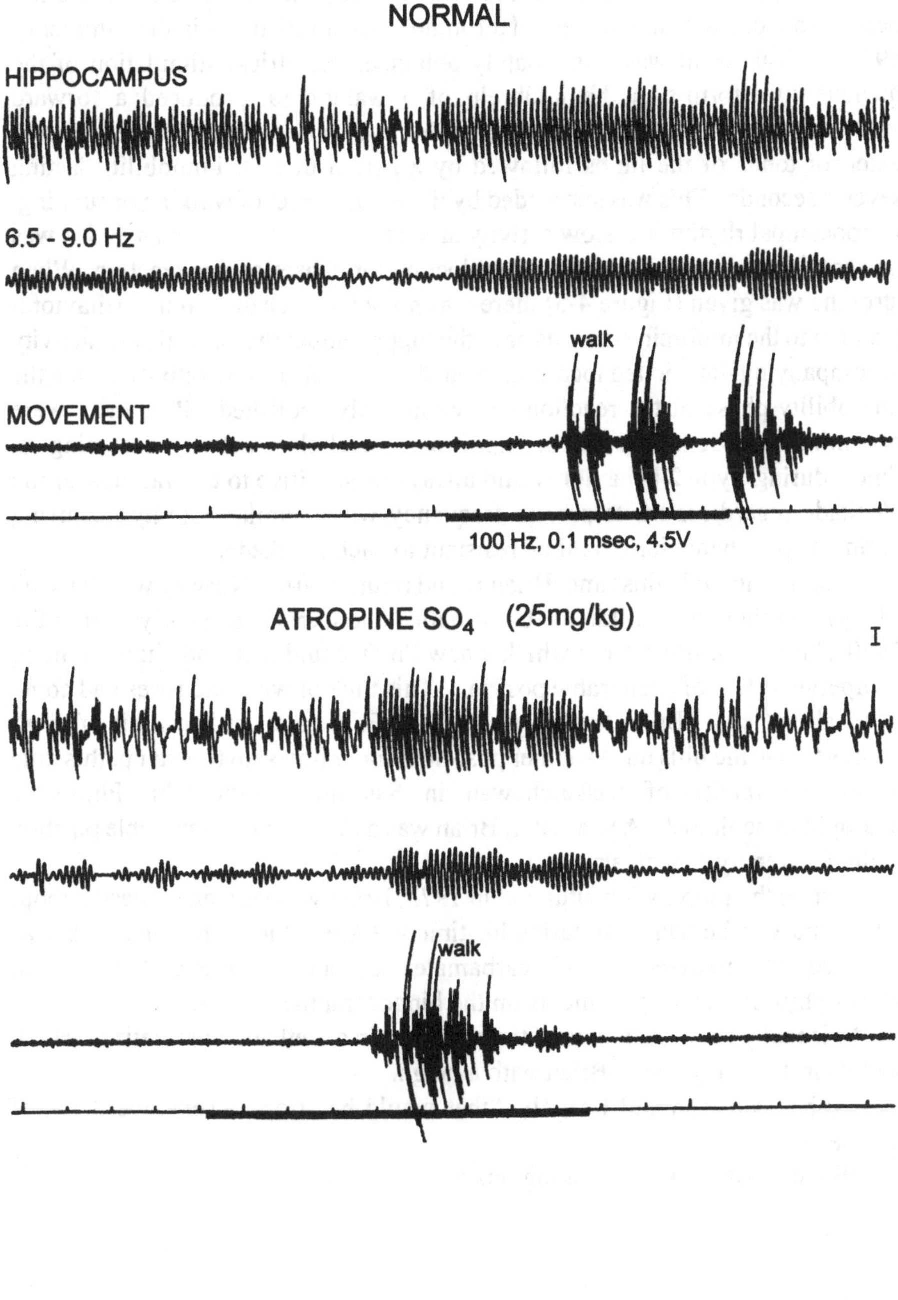

NORMAL
HIPPOCAMPUS
6.5 - 9.0 Hz
walk
MOVEMENT
100 Hz, 0.1 msec, 4.5V
ATROPINE SO_4 (25mg/kg)
I
walk

locomotion was provided by the effects of stimulating the midbrain reticular formation. Brian Bland had noticed that such stimulation could elicit hippocampal rhythmical slow activity in the immobile waking rat and study of this effect had provided an MSc. thesis for Craig Milne in 1972 but it did not occur to anyone to test the effect of atropine. When I tried this in the summer of 1973, a clear result was immediately obtained. Electrical stimulation of the nucleus cuneiformis in the midbrain of a waking rat produced a forward movement of the pinnae, widening of the eyes and a slight increase in the extensor tonus of the limbs followed by a period of total immobility lasting several seconds. This was succeeded by the abrupt onset of walking or running. Hippocampal rhythmical slow activity at a frequency of as much as 10 Hz was present during both the immobile and locomotor phases of this reaction. When atropine was given (Figure 4-5) there was no obvious change in the behavioral reaction to the midbrain stimulus or in the hippocampal rhythmical slow activity accompanying the elicited locomotion but the rhythmical slow activity during the immobility phase of the reaction was completely abolished. Ron and I were becoming confident that hippocampal rhythmical slow activity occurring, at times, during Type 2 behavior would always be sensitive to central muscarinic blockade regardless of its precise frequency while similar activity occurring during Type 1 behavior would be resistant to such blockade.

At approximately this time, Brian Bland returned from Norway with his wife Cheryl and their two children. Universities had expanded rapidly during the 1960s, creating a situation in which a new Ph.D. could pick and choose among competing offers of a tenurable position. This time of wine and roses had come to an abrupt end: Brian had no job offers. Fortunately, I had a bit of grant money to pay the bills until another post-doctoral fellowship turned up, this time at the University of Saskatchewan in Saskatoon with John Phillis, a neuropharmacologist. A year later, Brian was able to secure a tenurable position at the University of Calgary.

Despite the gloomy job situation in 1973, Brian was very enthusiastic about all the marvels he had seen during his time in Oslo. One of these marvels was the use of urethane (ethyl carbamate) as an anesthetic for use in electrophysiological experiments on the hippocampus.

"It is really easy to elicit good theta activity in a urethane-anesthetized rabbit, and it isn't moving," said Brian with a laugh.

"Ahh," said Ron and I sagely, "that would be atropine-sensitive theta, of course".

"What?" asked Brian, looking puzzled.

We soon found a supply of urethane and tried the effect of atropine. As expected, all the rhythmical slow activity that can be elicited in a urethanized rat or rabbit is readily blocked by atropine (Figure 4-6).

Larry Gillespie suggested that if atropine-sensitive and atropine resistant hippocampal rhythmical slow activity were truly dependent on neurochemically distinct pathways (i.e. dependent on different neurotransmitters) then they might mature at different rates in young animals. I thought this an excellent idea, so Larry began the ticklish business of implanting electrodes in infant rats, rabbits and guinea pigs. It turned out that young New Zealand White rabbits did not display rhythmical slow activity until they were 12-14 days of age, which is approximately the age at which the eyes and ears open and the rabbits first begin to hop about outside the maternal nest. This early rhythmical slow activity occurred only during the occurrence of Type 1 movement and it was resistant to atropine. Atropine-sensitive rhythmical slow waves, capable of being elicited during behavioral immobility or ether anesthesia did not appear until the age of 22-24 days. A similar result was obtained in rats: atropine-resistant Type 1 movement-related rhythmical slow activity appeared only at 12-14 days of age while rhythmical slow activity during ether anesthesia appeared at 20-22 days of age. These findings supported the idea that atropine-sensitive and atropine-resistant rhythmical slow activity are dependent on two distinct neural inputs to the hippocampus which mature at different rates. A different result was obtained in the precocial guinea pig, an animal which, unlike the altricial rats and rabbits, is perfectly capable of walking about within a minute or two after birth. Guinea pigs possessed both atropine-sensitive and atropine-resistant rhythmical slow activity on the day they were born.

Despite having completed a rather original piece of research which earned him a Ph.D. in 1975[9], Larry decided that research and the academic life in general were not to his liking. He chose not to publish his work, returned home to southern Alberta, and eventually became a cattle rancher.

A few years later, his experiments were partially repeated and extended by Brian Bland and two of his students using rats and Dutch Belted rabbits[10]. The results were similar to Larry's in many respects except that in the rats the drug eserine, an anticholinesterase inhibitor that enhances the actions of brain acetylcholine, produced hippocampal rhythmical slow activity during behavioral immobility as early as the eighth day of life. This indicates that a cholinergic input to the hippocampus develops at about the same time that the atropine-resistant movement-related input becomes active. Perhaps eserine is more potent than is ether anesthesia in activating the cholinergic pathway.

Figure 4-6. **Spontaneous or stimulus-elicited hippocampal theta activity during ether or urethane anesthesia in the rabbit and rat. Signal marker indicates application of stroking or pinching a foot. Following atropine sulfate (5 mg/kg, iv for rabbits and 50 mg/kg, ip for rats) all theta activity is abolished, and hippocampal activity consists of small amplitude irregular waves. From Kramis, R., Vanderwolf, C.H., and Bland, B.H. (1975). Two types of hippocampal rhythmical slow activity in both the rabbit and rat: relations to behavior and effects of atropine, diethyl ether, urethane, and pentobarbital.** ***Experimental Neurology, 49,*** **58-85.**

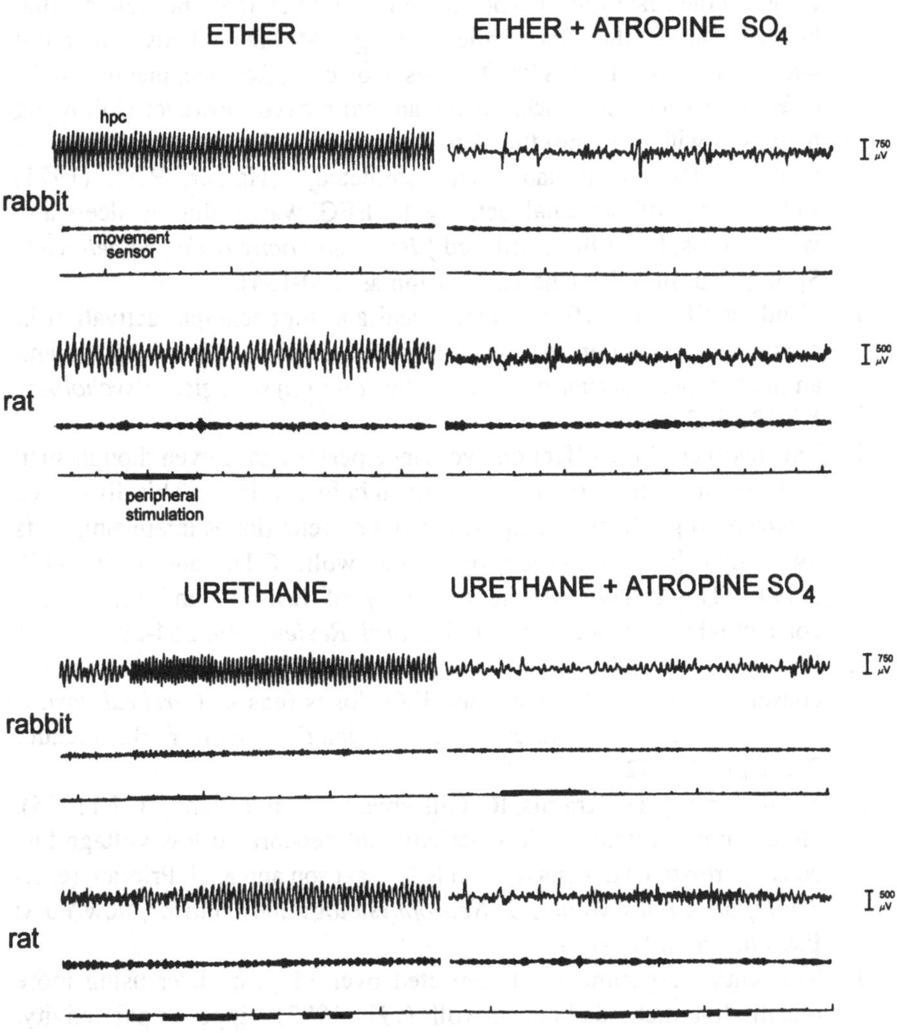
ETHER
ETHER + ATROPINE SO_4
hpc
rabbit
movement sensor
750 µV
rat
500 µV
peripheral stimulation
URETHANE
URETHANE + ATROPINE SO_4
rabbit
750 µV
rat
500 µV

Notes on Chapter 4

1. Penfield, W., and Jasper, H. (1954). *Epilepsy and the functional anatomy of the human brain.* Boston: Little, Brown & Co.
2. I was not the first to *see* this phenomenon. Ron Harper once told me that he had observed the effect while working at McMaster University but it was not mentioned in his Ph.D. thesis. However, Ron later mentioned the effect in a single sentence, "If the animal moved, however (following treatment with atropine) the slow waves would disappear and reappear after the movement had been completed," [Harper, R.M. (1973) Relationship of neuronal activity to EEG waves during sleep and wakefulness, In Phillips, M.I. (ed.) *Brain unit activity during behavior.* Springfield, Illinois: Charles C. Thomas, 130-154].
3. Vanderwolf, C.H. (1975). Neocortical and hippocampal activation in relation to behaviour: effects of atropine, eserine, phenothiazines and amphetamine. *Journal of Comparative and Physiological Psychology, 88*: 300-323.
4. This had very little effect on avoidance performance even though such performance is poor if the drug is given before training. This difference between drug effects on acquisition and on retention is interesting in its own right [see discussion by Vanderwolf, C.H., and Cain, D.P. (1994). The behavioural neurobiology of learning and memory: a conceptual reorientation. *Brain Research Reviews, 19*: 264-297]
5. Plum, F. (1991). Coma and related global disturbances of the human conscious state. In A. Peters and E.G. Jones (eds.). *Cerebral cortex, volume 9, Normal and altered states of function.* New York: Plenum Press, pp. 359-425.
6. Vanderwolf, C.H., Kramis, R., Gillespie, L.A., and Bland, B.H. (1975). Hippocampal rhythmic slow activity and neocortical low voltage fast activity: relations to behavior. In R.L. Isaacson and K.H. Pribram (eds.) *The hippocampus, volume 2: Neurophysiology and behavior*, New York: Plenum Press, 101-128.
7. This entire experiment was repeated over 15 years later using more quantitative methods [Vanderwolf, C.H. (1992) Hippocampal activity, olfaction and sniffing: an olfactory input to the dentate gyrus. *Brain Research, 593*: 197-208]. It was shown that although the atropine - or scopolamine-sensitive form of hippocampal rhythmical slow activity can precede the initiation of the jump response by several seconds, the atropine-resistant form of this activity precedes the initiation of jumping by only 100-200 milliseconds.

8. Kramis, R., Vanderwolf, C.H., and Bland, B.H. (1975). Two types of hippocampal rhythmical slow activity in both the rabbit and rat: relations to behavior and effects of atropine, diethyl ether, urethane, and pentobarbital. *Experimental Neurology, 49*: 58-85.
9. Gillespie, L.A. (1975). *Ontogeny of hippocampal electrical activity and behavior in rat, rabbit, and guinea pig.* Unpublished Ph.D. thesis, University of Western Ontario, London, Ontario, Canada.
10. Creery, B.L., and Bland, B.H. (1980). Ontogeny of fascia dentata electrical activity and motor behavior in the Dutch Belted rabbit (*Oryctolagus cuniculus*). *Experimental Neurology*, *67*: 554-572.
Leblanc, M.O., and Bland, B.H. (1979). Developmental aspects of hippocampal electrical activity and motor behavior in the rat. *Experimental Neurology*, *66*: 220-237.

Chapter 5

The Anatomical and Neurochemical Basis of Atropine-Resistant Neocortical Activation[1]

The experiments discussed in Chapter 4 suggested the existence of two distinct activating inputs to the neocortex. In 1973 it seemed reasonable to think that a cholinergic component of the ascending reticular activating system was responsible for atropine-sensitive activation while some ascending non-cholinergic system was responsible for atropine-resistant activation. Since there was evidence of various kinds to suggest the involvement of ascending noradrenergic fibres from the locus coeruleus in cortical activation, I suspected that atropine-resistant neocortical activation might be dependent on norepinephrine (noradrenalin).

As a first step in evaluating this hypothesis, I began studying the effects of reserpine on hippocampal and neocortical slow wave activity in relation to behavior. Reserpine, an alkaloid derived from an Indian vine (*Rauwolfia serpentina*), has played a major role in the development of neuropharmacology. Its main action is to damage the storage vesicles in central and peripheral aminergic neurons. The amines that are released from the storage vesicles are destroyed in the cytoplasm by the action of an enzyme, monoamine oxidase. Consequently, a large systemic dose of reserpine has very little immediate effect, but over a period of several hours there is a gradual depletion of adrenalin, noradrenalin, dopamine and serotonin from the central nervous system, the enteric nervous system, sympathetic neurons, and the adrenal glands. Associated with these depletions of central and peripheral aminergic neurotransmitters there

Figure 5-1. Neocortical slow wave activity in a rat following reserpine (10 mg/kg) plus atropine (50 mg/kg) or pimozide (5 mg/kg; an antagonist of some of the synaptic effects of dopamine) plus atropine (50 mg/kg). *Upper traces*: one hour after pimozide and 30 min after atropine, rat is severely cataleptic but makes occasional spontaneous head movements (h) which are associated with low voltage fast activity (LVFA). Walking and struggling are also associated with LVFA but large slow waves occur during immobility. *Lower traces*: same rat 24 days later. Forty-eight hours after reserpine and 30 min after atropine rat displays moderate spontaneous activity, moving the head, stepping and walking. Low voltage fast activity is absent. Time marks indicate 1- and 5-sec intervals. From Vanderwolf, C.H., and Pappas, B.A. (1980). Reserpine abolishes movement-correlated atropine-resistant neocortical low voltage fast activity. *Brain Research, 202*: 79-94 with the permission of Elsevier Science.

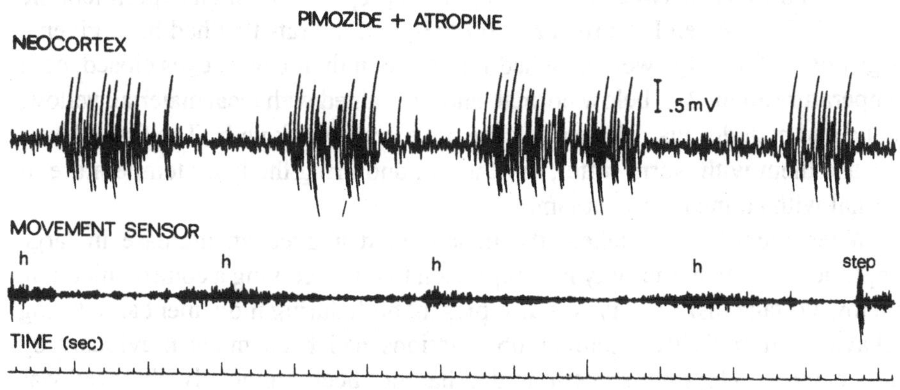
PIMOZIDE + ATROPINE
NEOCORTEX
.5 mV
MOVEMENT SENSOR
h
h
h
h
step
TIME (sec)

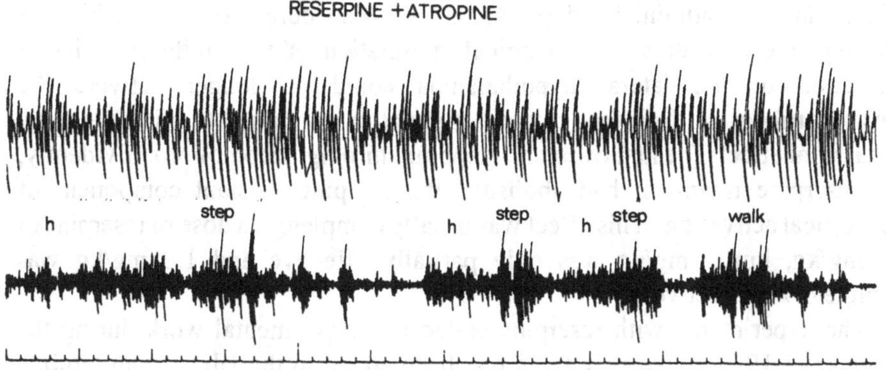
RESERPINE + ATROPINE
h
step
h
step
h
step
walk

is a reduction in the size of the pupil (miosis) together with a partial or complete closure of the eye (ptosis), a mild hypotension, a marked increase in peristaltic activity (all due, presumably, to a reduction in sympathetic activity), a fall in core temperature and a marked reduction in spontaneous motor activity (akinesia). If treated rats are placed in unnatural positions, for example, with one forepaw placed on top of a large rubber stopper, they make no move to return to a normal posture (catalepsy).

The method I adopted was to inject rats with reserpine before I went home at the end of the day (between 6:00-9:00 pm), in preparation for an experiment the next morning. When I returned at 8:00-8:30 am, the rats that had been given a large dose (10 mg/kg) were hunched immobile in their cages, eyes closed, core temperature about 2^0 C below normal, and fur stained with fecal material, yellow, liquid and malodorous. These rats were not a joy to work with. The first job was to wash them with warm water, dry the fur, and bring the body temperature to normal with an incandescent lamp.

When records were taken, the neocortex displayed an increase in large amplitude slow wave activity in comparison to rats receiving a control injection, but low voltage fast activity was also present both during movement and during behavioral immobility. Similar observations had been made previously by earlier investigators but something that had not been previously observed was that all neocortical low voltage fast activity disappeared when an adequate dose of atropine was administered (Figure 5-1). Procedures such as pushing or pinching the rats, or strong electrical stimulation of the midbrain reticular formation or the lateral hypothalamus could no longer activate the electrocorticogram even though such stimulation continued to elicit vigorous behavioral effects (head movement, stepping, turning, locomotion). Evidently, the reserpine treatment had abolished the atropine-resistant component of neocortical activation. This effect was usually complete at a dose of reserpine of 10 mg/kg, but 5 mg/kg was only partially effective and 1.0 mg/kg was completely ineffective.

The experiments with reserpine ended my experimental work during the summer of 1973. Two new faces had been added to the lab. Bryan Kolb, a former Calgarian who had just completed a Ph.D. with J.M. Warren at Pennsylvania State University arrived in June to take up a position as a post-doctoral fellow for one year. Terry Robinson who had just completed a master's degree with Ian Whishaw and Tom Wishart (a graduate of the University of Western Ontario who had taken a post at the University of Saskatchewan in Saskatoon) arrived in September to begin work on a Ph.D. degree.

We had a new phenomenon to work on. Since atropine-resistant activation of the neocortex was sensitive to reserpine, it seemed very likely that it was dependent on central norepinephrine, dopamine, or serotonin. There were two obvious strategies to determine which of these amines was the essential one (if indeed, one of them was essential). One strategy was to eliminate the effect of these amines selectively by means of drugs that block selectively the synthesis or the post-synaptic actions of each amine or by localized destruction of the relevant neurons. The other approach was an adaptation of an experiment first performed by A. Carlsson and his colleagues in Sweden. These investigators discovered that treatment with 3,4-dihydroxyphenylalanine (L-DOPA, the biochemical precursor of dopamine and norepinephrine) restored spontaneous motor activity in rats or rabbits pretreated with reserpine but that treatment with 5-hydroxtryptophan (the immediate precursor of serotonin) had no effect. This was the first piece of clear evidence that dopamine or norepinephrine played an essential role in locomotor ability but that serotonin had no such role.

Work in this field demanded an ability to determine the effects of the experimental treatments on the levels of brain amines and their metabolites. Since we had no facilities to do this at Western, I was fortunate to secure the collaboration first of Bruce Pappas at Carleton University in Ottawa, who made use of fluorometric techniques, and later of Glen Baker at the University of Alberta in Edmonton, who made use of high pressure liquid chromatography. It became a frequent occurrence for one of us (usually Richard Cooley) to rush to the airport with a large box containing dissected parts of rat brains frozen and packed in solid carbon dioxide. A few weeks later, letters or faxes containing tables of numbers would appear, as if by magic.

Numerous experiments on the effect of selective elimination of dopamine and noradrenalin (catecholamines) were carried out in the next few years by Terry Robinson and me and by Bryan Kolb (who had moved to the University of Lethbridge) and Ian Whishaw. Antagonists such as various neuroleptic or antipsychotic drugs (to block the synaptic effects of dopamine) or anti-adrenergic drugs, such as phenoxybenzamine or propranolol, appeared to have no effect, even in large doses, on atropine-resistant low voltage fast activity in the neocortex. A typical result of such an experiment is shown in Figure 5-1. Blockade of the synthesis of the catecholamines with α-methyl-*p*-tyrosine or surgical destruction of the locus coeruleus (the location of most of the noradrenergic neurons that project to the forebrain) was also without direct effect. Surgical lesions of the zona compacta of the substantia nigra, destroying the ascending dopaminergic neurons, had no effect either. Injection of 6-hydroxydopamine (a specific neurotoxin for catecholaminergic neurons) into the cerebral ventricles depleted brain dopamine and noradrenalin by as much as 99 percent without eliminating atropine-resistant neocortical and hippocampal activation.

Surgical lesions of the lateral hypothalamus, however, produced a loss of atropine-resistant neocortical activation, suggesting that this type of activation was dependent on an ascending pathway running through the lateral hypothalamus. This was interesting but did not allow a distinction to be made between ascending noradrenergic, serotonergic, or dopominergic pathways since all of them run through the medial forebrain bundle in the lateral hypothalamic region.

It should be noted that many of these treatments had *indirect* effects on atropine-resistant activation of the neocortex because they decreased or increased the occurrence of spontaneous head movements and locomotion. Thus, after treatment with neuroleptic drugs (blocking dopaminergic synapses), atropine-resistant neocortical activation occurred less frequently than before but its association with Type 1 movement persisted. Conversely, increasing the release of dopamine and noradrenalin by administration of *d*-amphetamine greatly increased the occurrence of atropine-resistant neocortical activation in correlation with a large increase in walking, rearing, and head movement. Therefore, the results appeared to indicate that the catecholamines were not *directly* involved in atropine-resistant activation of the neocortex, but that they had a major *indirect* role.

It seemed unlikely to me that serotonin would have a role in activation of the cerebral cortex: the dominant view in the 1970s was that serotonin promoted sleep. Further, we found that methysergide bimaleate, said at the time to block most of the synaptic effects of serotonin, did not block atropine-resistant activation of the neocortex. A single 500 mg/kg dose of *p*-chlorophenylalanine, an effective inhibitor of the synthesis of serotonin, was also ineffective as Terry Robinson was the first to show. Therefore, it seemed safe to conclude that atropine-resistant neocortical low voltage fast activity could not possibly be dependent on serotonin.

The results of the selective amine replacement experiments seemed at first, to corroborate the conclusions suggested by the selective amine elimination experiments. 5-Hydroxytryptophan, the immediate precursor of serotonin, given to rats pretreated with reserpine, did not restore atropine-resistant activation. L-DOPA, the immediate precursor of dopamine, produced a very dramatic restoration of spontaneous motor activity in reserpinized rats but did not restore atropine-resistant cortical activation. It was most impressive to watch a hyperactive rat treated with reserpine, benserazide, L-DOPA and atropine, walking, rearing, and biting ferociously at the apparatus or at the experimenter's glove while all the while neocortical activity consisted of large slow waves of the kind normally seen in quiet sleep. Similar effects were observed if amphetamine was given instead of L-DOPA and benserazide. These experiments seemed to rule out any possibility that atropine-resistant neocortical activation was dependent on the catecholamines or serotonin. Numerous other compounds were tested for their ability to restore atropine-resistant neocortical activation in reserpinized rats. Nothing worked.

By 1978 it seemed that this whole line of research was going nowhere. All the results we had were negative; the principal known pharmacological effects of reserpine appeared to be unrelated to its effects on atropine-resistant activation. I began to look for a way of salvaging something: perhaps I could demonstrate that reserpine had an important effect on cerebral activity that had nothing to do with dopamine, noradrenalin, or serotonin. This was important to me because the requirements of granting agencies demand a steady flow of publications. One must keep up an appearance of "progress" even when nothing of significance is being discovered.

I reasoned as follows. The depleting effect of reserpine on brain amines is dependent on the action of the enzyme monoamine oxidase. If I were to block the activity of monoamine oxidase before adminstering reserpine, brain amines would be protected. If atropine-resistant activation disappeared despite this, then one could conclude that reserpine was acting by a mechanism distinct from its effect on brain amines.

I first tried this experiment using nialamide, an effective inhibitor of brain monoamine oxidase. The results of the first day of work were very clear. Nialamide completely protected atropine-resistant neocortical low voltage fast activity from the destructive effects of reserpine.

In the early evening I went off to have supper by myself at the Great Hall, the main campus cafeteria at the time, thinking all the while that this whole endeavour was a complete mess. Food and coffee revived my spirits however, giving the results a somewhat rosier appearance. The nialamide experiment indicated that a mystery substance (or substances) involved in atropine-resistant activation could be inactivated by monoamine oxidase. Therefore, it was probably a monoamine. Nonetheless the previous research had eliminated, or so it seemed, any possibility that the catecholamines or serotonin were essential for atropine-resistant activation. Therefore, there must be other amines in the brain which are substrates for monoamine oxidase. I said to myself, "This isn't a mess. It's a discovery!"

I spent the next day in the library reading everything I could find on monoamine oxidase. I discovered that there was a group of substances in the brain known to biochemists as trace amines owing to their extremely low concentration under normal conditions. These compounds were all readily inactivated by monoamine oxidase. The most frequently studied member of the group, a substance known as β-phenylethylamine, would enter the brain readily when given by the usual systemic routes of injection. Immediately, I ordered a supply.

When the replacement experiment was repeated a few weeks later, using β-phenylethylamine, the result was dramatic. β-Phenylethylamine rapidly restored low voltage fast activity in the neocortex in rats pretreated with reserpine plus atropine. It also had a peculiar behavioral effect consisting mainly of the elicitation of side-to-side movements of the head and alternating flexion-extension (pawing movements) of the forelimbs. I was quite unaware of the significance of this, but I was ecstatic about the restoration of atropine - resistant neocortical activation. At last I had a clear positive result![2] I got everyone around the lab that day to come and look at it. Sober second thought however, soon dispelled any idea that β-phenylethylamine was the substance, acted on by reserpine, which normally produced atropine-resistant activation. For one thing, Allan Boulton and other investigators at the University of Saskatchewan had shown in 1977 that reserpine did not reduce the levels of β-phenylethylamine in the brain.

Since nialamide protected atropine-resistant neocortical activation from the effects of a subsequent treatment with reserpine, it was natural to wonder whether it would also restore atropine-resistant activation when it had already been abolished by prior treatment with reserpine. Nialamide had this effect, as did other monoamine oxidase inhibitors (pargyline, tranylcypromine) and the effect occurred within 15-30 min (Figure 5-2).

Shortly after this, I attended a paper session devoted to trace amines at the annual meeting of the Society for Neuroscience. I was surprised to learn that the behavioral syndrome elicited by β-phenylethylamine resembled the "serotonin behavioral syndrome" produced by a variety of pharmacological treatments that had in common the fact that they all released serotonin in the central nervous system. It had been established that the peculiar pawing movements and the side-to-side movements of the head were due to activation of descending serotonergic pathways from the lower brain stem to the spinal cord. Consequently β-phenylethylamine might be acting by releasing residual serotonin or by mimicking the action of serotonin at some of its receptors both in the brain and in the spinal cord.

I had never paid much attention to work on serotonin since I had no reason to doubt the conclusion, frequently voiced at that time, that it acted to promote sleep. Apart from this, I thought I had three lines of evidence that serotonin could not be the amine involved in atropine-resistant neocortical activation. (1) The serotonin antagonists methysergide bimaleate, cyproheptadine hydrochloride, metergoline and spiperone (spiroperidol) did not block atropine-resistant neocortical activation, even in very large doses. (2) Treatment with 5-hydroxytryptophan, the immediate precursor of serotonin, did not restore atropine-resistant neocortical activation in reserpinized rats. (3) A single large

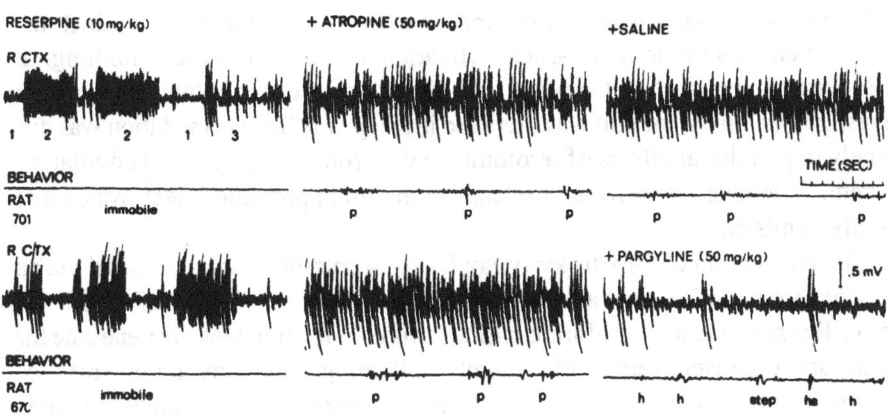

Figure 5-2. Pargyline-induced restoration of atropine-resistant low voltage fast activity in a reserpinized rat. *Upper traces*: Control rat (#701). After reserpine, electrocorticogram includes low voltage fast activity (1), rhythmical spindles (2) and large irregular slow waves (3) all occurring during behavioral immobility. After atropine, all low voltage fast activity is abolished, even during push trials (P) in which the rat was pushed forward to elicit stepping or struggling movements. No further change after saline administration. *Lower traces*: Experimental rat (#670) resembles control rat after reserpine and reserpine + atropine but displays prominent low voltage fast activity during spontaneous behavior after administration of pargyline. H, head movement; step, stepping; hs, head shake. From Vanderwolf, C.H. (1984). Aminergic control of the electrocorticogram: a progress report. In A.A. Boulton, G.B. Baker, W.G. Dewhurst, and M. Sandler (eds.) *Neurobiology of the trace amines*. Clifton, New Jersey, Humana Press, pp. 163-183 with the permission of Humana Press.

dose of *p*-chlorophenylalanine, which blocks the synthesis of serotonin, did not abolish atropine-resistant neocortical activation.

However from what I had learned from time in the library and from talking to various people interested in serotonin, I became aware that: (a) serotonin antagonists that work well in peripheral tissues are often completely ineffective in the brain; and (b) most of an injected dose of 5-hydroxytryptophan is converted to serotonin in peripheral tissues or in the walls of capillaries in the brain: it never reaches brain neurons at all. Serotonin itself penetrates the brain very poorly. Consequently, my first two lines of "evidence" against the involvement of serotonin in cortical activation amounted to exactly nothing. It also became apparent that much of the previous evidence, obtained by other investigators, that serotonin was involved in sleep or behavioral sedation was due largely to peripheral effects of serotonin on the bronchioles, gut and circulatory system. The idea that serotonin release in the brain promotes sleep was, quite simply, a mistake.

The conversion of 5-hydroxytryptophan to serotonin is accomplished via an enzyme, aromatic L-amino acid decarboxylase, which is widely distributed in the body. Benserazide hydrochloride inhibits this enzyme but does not penetrate the blood-brain barrier very well, thereby allowing the conversion from 5-hydroxytryptophan to serotonin to occur in the central nervous system but not in peripheral tissues. Consequently an injected dose of 5-hydroxytryptophan given shortly after treatment with benserazide is converted to serotonin in the brain but the complicating effects of high peripheral levels of serotonin are eliminated. When this drug combination was given to reserpinized rats, atropine-resistant neocortical low voltage fast activity was rapidly restored. Administration of benserazide plus tyrosine, benserazide plus phenylalanine or benserazide plus L-DOPA did not have this effect. This was the first clear cut evidence we had that the mystery amine involved in atropine-resistant neocortical activation was really serotonin.

If the amine essential to atropine-resistant neocortical activation is really serotonin, treatment with *p*-chlorophenylalanine should have the same effect as reserpine, producing a loss of this form of activation. Re-examination of this problem revealed that a single 500 mg/kg dose of *p*-chlorophenylalanine, which did not abolish atropine-resistant neocortical activation, produced a decline of brain serotonin to 10.6% of control levels and a decline of 5-hydroxyindoleacetic acid (the main metabolite of serotonin) to 10.5% of control levels. Three successive daily doses of *p*-chlorophenylalanine reduced these levels to 8.0% and 6% respectively. This difference is rather small but it seemed to have large functional consequences. Three daily doses of *p*-chlorophenylalanine usually resulted in a total disappearance of atropine-resistant neocortical activation (see Figure 5-3).

Large doses of *p*-chlorophenylalanine undoubtedly have a variety of non-specific effects: it would be highly desirable to confirm the results by some more specific method. Fortunately such a method was available. Intracerebral injections of 5,7-dihydroxytryptamine, a neurotoxin with considerable specificity for serotonergic neurons, abolished atropine-resistant neocortical activation in much the same way as systemic injections of reserpine or *p*-chlorophenylalanine[3].

Therefore, we had converging evidence from several methods that the mystery amine involved in atropine-resistant neocortical activation really is serotonin. However, the fact that putative serotonin antagonists failed to abolish atropine-resistant neocortical activation in freely moving rats continued to pose a problem. Over a period of several years, 13 different antagonists were tested. All of them were completely ineffective except for methiothepin maleate which was only partially effective, even in rather large doses[4]. The nature of the neocortical serotonin receptors involved in atropine-resistant neocortical activation is still an unsolved problem.

Another test of the hypothesis that serotonin is essential for the occurrence of atropine-resistant neocortical activation consisted of electrical or pharmacological stimulation of serotonergic pathways. Terry Robinson and, a decade later, Barbara Peck, a student who came from Queen's University in Kingston, Ontario to do graduate work in 1988, showed that electrical stimulation of the dorsal raphe nucleus and some sites in the median raphe nucleus (the places where most of the serotonergic cell bodies that send axons to the cerebral cortex are located) produced locomotion accompanied by atropine-resistant neocortical activation. Barbara Robertson, another graduate student who began her career at Western in 1986, after receiving a BSc. at the University of Alberta in Edmonton, showed that drug treatments that promote serotonin release in the brain (tranylcypromine plus tryptophan, *p*-chloroamphetamine) also produce atropine-resistant neocortical activation.

All of this work shows beyond reasonable doubt, I think, that serotonin really is an essential neurotransmitter in the production of atropine-resistant neocortical activation. I feel rather embarrassed by the fact that it took 10 years to come to this conclusion and an additional 5 years to gather all the evidence. Part of the problem lay in having accepted false information as the truth. For a long time I believed that such drugs as methysergide, metergoline, or spiperone or a single 300-500 mg/kg dose of *p*-chlorophenylalanine were sufficient to block serotonergic activity in the brain and that an injected dose of 5-hydroxytryptophan would, by itself, increase serotonergic activity in the brain. All of these "facts" turned out to be untrue.

Research is sometimes compared to putting together a jig-saw puzzle. If this analogy is to be accepted, it is important to understand that in the research puzzle

Figure 5-3. Effects of *p*-chlorophenylalanine and scopolamine on the electrocorticogram in relation to behavior. *Cortex*, slow wave activity from parietal neocortex, surface-to-depth electrodes. *Movement*, output from magnet-and-coil type of movement sensor. *Vehicle, scopolamine*; rat treated with 3 daily injections of drug vehicle plus scopolamine (5 mg/kg., SC). *PCPA, scopolamine*; rat treated with 3 daily injections of *p*-chlorophenylalanine (500 mg/kg, IP) plus scopolamine (5 mg/kg, SC); im, immobile; w, walking. Note that in the PCPA-scopolamine rat large amplitude irregular slow waves occur continuously, but that in the vehicle-scopolamine rat such waves occur only during behavioral immobility. Cal.: 1.0 mV, 5 sec. From Vanderwolf, C.H., Dickson, C.T., and Baker, G.B (1990). Effects of *p*-chlorophenylalanine and scopolamine on retention of habits in rats. *Pharmacology, Biochemistry and Behavior, 35*: 847-853, with permission from Elsevier Science.

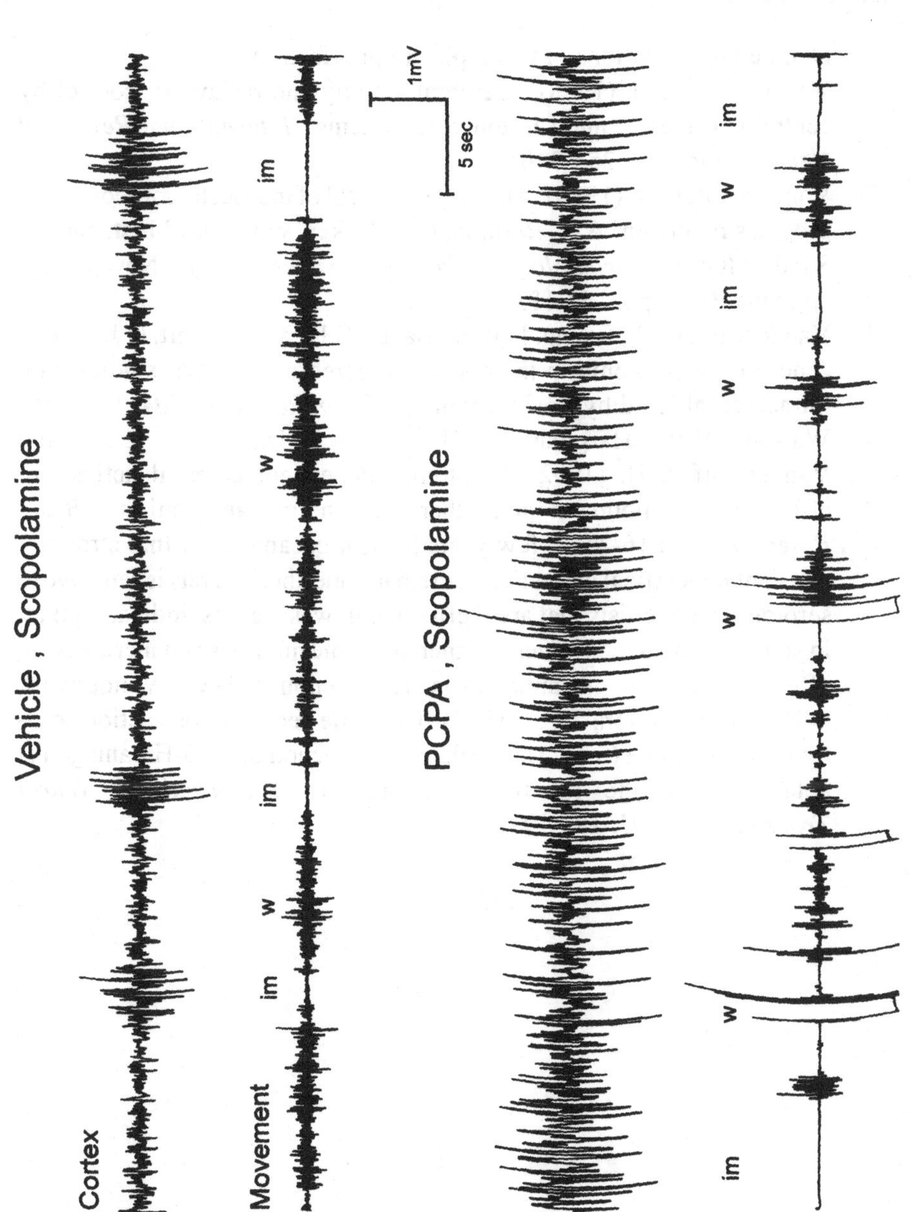
Vehicle , Scopolamine
Cortex
Movement
im
w
im
w
im
1mV
5 sec
PCPA , Scopolamine
im
w
w
w
im
w
im

the number of pieces is unknown, many of the pieces are missing or belong to a different puzzle, some of the pieces are simply random bits of material that do not form a part of any puzzle, and finally, one has no idea what the final picture will look like.

Notes on Chapter 5

1. Detailed presentations of this topic are provided by: Vanderwolf, C.H. (1988). Cerebral activity and behavior: control by central cholinergic and serotonergic systems. *International Review of Neurobiology*, *30*: 255-340.
2. Vanderwolf, C.H. (1984). Aminergic control of the electrocorticogram: a progress report. In: A.A. Boulton, G.B. Baker, W.G. Dewhurst, and M. Sandler (eds.) *Neurobiology of the trace amines*. Clifton, New Jersey, Humana Press, pp. 163-183.
3. Vanderwolf, C.H., Leung, L.W.S., Baker, G.B., and Stewart, D.J. (1989). The role of serotonin in the control of cerebral activity: studies with intracerebral 5,7-dihydroxytryptamine. *Brain Research*, *504*: 181-191.
4. Watson, N.V., Hargreaves, E.L., Penava, D., Eckel, L.A., and Vanderwolf, C.H. (1992). Serotonin- dependent cerebral activation: effects of methiothepin and other serotonergic antagonists. *Brain Research*, *597*: 16-23. A few years later, it became clear that atropine-resistant neocortical activation in urethane anesthetized rats is sensitive to serotonin antagonists that are ineffective in waking rats, indicating that, in some as yet undetermined manner, urethane modifies and increases the effect of serotonergic antagonists. [Dringenburg, H.C., Vanderwolf, C.H., and Hamilton, J.T. (1995). Urethane reduces contraction to 5-hydroxytrytamine (5-HT) and enhances the action of the 5-HT antagonist ketanserin on the rat thoracic aortic ring. *Journal of Neural Transmission*, *101*, 183-193].

Chapter 6

The Anatomical and Neurochemical Basis of Atropine-Resistant Hippocampal Rhythmical Slow Activity

Hippocampal slow wave activity was recorded together with neocortical activity in many of the experiments described in Chapter 5. In most respects, atropine-resistant hippocampal rhythmical slow activity appeared to operate in parallel with atropine-resistant neocortical activation. Thus, like the corresponding neocortical wave pattern, atropine-resistant hippocampal rhythmical slow activity was not directly affected by dopaminergic or noradrenergic synaptic blockade, by lesions of the substantia nigra or the locus coeruleus, by blockade of the synthesis of catecholamines by α-methyl *p*-tyrosine, by intraventricular injection of 6-hydroxydopamine or by such serotonergic antagonists as methysergide or metergoline. Neuroleptic drugs (blocking dopamine receptors) produced a striking reduction in the occurrence of atropine-resistant hippocampal rhythmical slow activity but did not alter its correlation with Type 1 movement: the movement and the hippocampal wave form always occurred together even though both occurred much less frequently than under normal conditions. Conversely, the occurrence of the atropine-resistant hippocampal waveform and the associated Type 1 behavior (mainly walking, rearing, and head movement) were greatly increased by the administration of a moderate dose of *d*-amphetamine. These observations suggest that atropine-resistant inputs to both the hippocampus and the neocortex are brought into action by a dopaminergic mechanism.

Terry Robinson carried out studies of the effect on behavior and neocortical and hippocampal activity of electrical stimulation of 129 sites ranging from the rostral midbrain to the caudal medulla[1]. Stimulation at most sites elicited such motor effects as head movement, walking, running, or turning in a tight circle. These reactions were always associated with hippocampal rhythmical slow activity which was not abolished by administration of atropine sulfate. Atropine-resistant neocortical activation was also elicited. Stimulation at only 4/129 sites elicited movement in the absence of hippocampal rhythmical slow activity in the undrugged state. These sites were in the cerebral peduncle (producing leaning to one side, rearing and contralateral limb movements), the facial nucleus and the trigeminal motor nucleus (producing closure of the ipsilateral eye and chewing, respectively). Since all these sites are in classical efferent pathways with no known ascending fibers, it is perhaps to be expected that activation of the hippocampus was not observed. From the point of view of brain function, movements elicited in this way are comparable to passive movements, which also fail to activate the hippocampus.

Stimulation of a variety of brain stem sites produced hippocampal rhythmical slow activity during complete immobility. With one exception (stimulation in the vicinity of the subnucleus compactus nuclei pedunculopontini tegmenti) this type of activity was completely abolished by the administration of atropine sulfate. Among a great variety of experimental procedures, this is the only one ever found (short of peripheral neuromuscular blockade by curare-like drugs) that would permit atropine-resistant hippocampal rhythmical slow activity in the absence of Type 1 movement. There is a comparable zone for the neocortex in the region of the nucleus of the posterior commissure and the nucleus cuneiformis. Stimulation in this zone in an atropinized rat will activate the neocortex during complete immobility. Stimulation at a great many other brainstem sites will activate the neocortex in an undrugged immobile rat but this reaction is invariably abolished by atropine administration.

These observations pose an interesting problem which no one has ever followed up. Perhaps stimulation in the nucleus compactus and the nucleus cuneiformis activate simultaneously ascending pathways that promote Type 1 movement and descending pathways that inhibit movement. The behavioral reaction at both sites consists of a brief period of immobility ("freezing") followed by a violent burst of running or jumping. If the rat is moving spontaneously when stimulation is applied, the movement always stops abruptly, lending some credence to the idea that movement is actively inhibited.

The reserpine experiments, which had worked well in the case of the neocortex, were of little value in studying atropine-resistant hippocampal rhythmical slow activity. Reserpine altered the atropine-resistant hippocampal wave form but did not abolish it altogether, at least not at a dose of 10 mg/kg. Consequently, this line of work was not pursued[2].

Terry Robinson successfully defended his Ph.D. thesis in November 1977 and spent the next year working as a post-doctoral fellow with Gary Lynch at the University of California at Irvine, California. Subsequently, he was able to obtain a tenurable position at the University of Michigan in Ann Arbor, Michigan.

In September, 1978, Lai-Wo (Stan) Leung arrived to spend a year at Western as a post-doctoral fellow. Stan had just completed a Ph.D. with Walter Freeman at the University of California at Berkeley, California. Subsequently, he spent another year as a post-doctoral fellow with Fernando Lopes da Silva at the University of Amsterdam, before returning to Western in 1980 on a special five year post-doctoral award from the Natural Sciences and Engineering Research Council. Finally, he found himself a tenurable position in the Department of Physiology at Western, becoming a full professor a few years later. Consequently, I had the good luck to work with Stan over a period of many years.

One of the problems that Stan and I worked on together was the role of the entorhinal cortex in the control of hippocampal slow wave activity. Anatomical studies had shown, many years before, that the hippocampus receives afferent fibres from both the septal region and the entorhinal cortex. It had been shown in the 1950s that the septal nuclei exerted a strong control over hippocampal slow waves but the role of the entorhinal cortex in this respect, if any, had not been established.

Stan and I investigated the effect of large unilateral and bilateral surgical removals of the entorhinal cortex on hippocampal slow wave activity[3]. Stan's academic background was largely in biophysics. Consequently, he was very familiar with the use of computers to perform analyses of slow wave, evoked potential or unit data, techniques which were quite unfamiliar to me. This added a valuable quantitative dimension to the work.

We found that hippocampal slow wave activity was surprisingly normal in rats with extensive bilateral removals of entorhinal cortex. During walking or struggling, rhythmical slow activity of about 7-8 Hz was present, as it is in normal rats. During isolated head movements the frequency was significantly lower (about 6.5 Hz), also as it is in normal rats. Although there were a number of minor electrophysiological abnormalities in the hippocampus in rats with entorhinal lesions, the most significant finding was that all rhythmical slow activity disappeared when atropine sulfate was administered. On the other hand, good rhythmical slow activity could be elicited under urethane anesthesia in rats that had received chronic entorhinal lesions. These findings indicated that the input that produced atropine-resistant hippocampal rhythmical slow activity reached the hippocampus via the entorhinal cortex. Presumably, the atropine-sensitive type of hippocampal rhythmical slow activity that was present during urethane anesthesia and during Type I behaviour in a rat with entorhinal cortex lesions was produced by a muscarinic cholinergic input via the septal nuclei.

A few years prior to these experiments, Ian Whishaw and Bryan Kolb had shown that lesions of the lateral hypothalamic area temporarily abolished both atropine-resistant hippocampal rhythmical slow activity and atropine-resistant neocortical activation[4]. The new data on the effect of entorhinal cortex lesions suggested that the atropine-resistant input to the hippocampus must run through both the lateral hypothalamic area and the entorhinal cortex.

Stan and I embarked on an extensive series of tract-cutting experiments to determine the course of these fibers. We found, first of all, that making an extensive cut, in the coronal plane, through the cingulate cortex and the entire neocortex, right down to the rhinal fissure, had little effect on hippocampal rhythmical slow activity in the waking undrugged rat and did not alter its normal correlation with motor activity, but it completely eliminated all traces of atropine-resistant hippocampal rhythmical slow activity in the freely moving rat. The rhythmical slow activity occurring during urethane anesthesia was unaffected (it was, of course, sensitive to atropine). Partial lesions showed that the main component of the pathway ran through the cingulate cortex, perhaps in the cingulum or the supra-callosal stria. Consequently, the fibers mediating atropine-resistant hippocampal rhythmical slow activity appeared to run through the lateral hypothalamus, up around the genu of the corpus callosum, then back through the cingulum and supra-callosal fibers to the entorhinal region[5].

In the summer of 1981, Gyorgy Buzsaki, a young neuroscientist on leave of absence from the University of Pecs in Hungary, arrived in the lab to take up a position as a post-doctoral fellow. I had met Gyorgy at a meeting on "*Neural plasticity and memory formation*" which was held in 1980 in Magdeburg in what was then known in the West as Communist East Germany. Gyorgy, being very eager to escape from the Communist world, was attempting to find a permanent position for himself somewhere in the West. He first spent the year 1980-81 working with Eduardo Eidelberg at the University of Texas in San Antonio, then spent a second year with us at Western.

Up to this point I had been interested mainly in two general problems: (1) to work out how the electrocortical activation patterns of the hippocampus and neocortex were related to behavior; and (2) to identify the essential neurochemical and anatomical systems involved in these activation patterns. Both Gyorgy and Stan were interested in a third problem: what are the detailed cellular mechanisms that produce rhythmical slow waves and other large-scale electrophysiological phenomena in the hippocampus?

Gyorgy was determined to make the most of his time. He worked long hours, setting up apparatus to record single units as well as slow waves in the hippocampus of freely moving rats, doing surgery, and taking records, often moving at a dog-trot from one task to another. A lengthy paper published in 1983[6] provided extensive experimental support for a neuronal model which proposed that hippocampal rhythmical slow activity was produced by a feed-forward inhibitory input from the septal nuclei to both Ammon's horn pyramidal cells and dentate granule cells, plus an excitatory input from the entorhinal cortex to the same classes of neurons.

After his two years in North America, Gyorgy was obliged to return to Hungary but he maintained a close contact with us and with other investigators in the United States. Gyorgy was the prime mover in arranging an international conference on the electrophysiology of the hippocampus held in Pecs in 1984 to commemorate the 60th birthday of his old mentor Endre Grastyan[7]. Eventually, Gyorgy's hopes were fulfilled: he obtained a tenured position at the center for Molecular and Behavioral Neuroscience at Rutgers University in Newark, New Jersey.

By the early 1980s we were slowly coming to the conclusion that the entorhinal input involved in the production of atropine-resistant hippocampal rhythmical slow activity might be dependent on serotonin. Serotonergic projections from the brain stem were known to run forward through the lateral hypothalamus, around the genu of the callosum and through the supracallosal stria and the cingulum to reach the entorhinal region and the hippocampus. This was the pathway identified by the tract-cutting experiments Stan Leung and I had done. Furthermore, the experiments on the neocortex going on at this time were showing that atropine-resistant activation of that structure is dependent on serotonin. It soon became apparent that atropine-resistant hippocampal rhythmical slow activity could be abolished by three daily doses of *p*-chlorophenylalanine[8]. A bit later we discovered further that intra-brainstem injections of 5,7-dihydroxytryptamine also eliminated atropine-resistant hippocampal rhythmical slow activity (Figures 6-1 and 6-2). We had also an abundance of evidence that the noradrenergic projections from the locus coerulus were not essential to atropine-resistant hippocampal rhythmical slow activity. Therefore, the long-sought mystery amine responsible for the noncholinergic activation of both the hippocampus and the neocortex turned out to be none other than serotonin.

Notes on Chapter 6

1. Robinson, T.E., and Vanderwolf, C.H. (1978). Electrical stimulation of the brain stem in freely moving rats: II. Effects on hippocampal and

Figure 6-1. Hippocampal slow wave activity in a freely moving rat that had received 4 intrabrainstem injections of Locke's solution (1.0 µl in each of: the dorsal raphe nucleus, the median raphe nucleus, and bilaterally in the region dorsal to the medial lemniscus). HIPP, hippocampus; 6-12 Hz, integrated 6-12 Hz hippocampal activity: MVMT, output from movement sensor. A: no drug. B: after treatment with scopolamine (5 mg/kg, s.c.). In both A and B note the presence of rhythmical slow activity (RSA) and the resulting increase in 6-12 Hz activity during walking as compared to immobility. Rat 2261. Calibration: 1 s, 1 mV. Figures 6-1 and 6-2 are taken from Vanderwolf, C.H., Leung, L.W.S., Baker, G.B., and Stewart, D.J. (1989). The role of serotonin in the central of cerebral activity: Studies with intracerebral 5,7-dihydroxytryptamine. *Brain Research, 504*: 181-191, with the permission of Elsevier Science

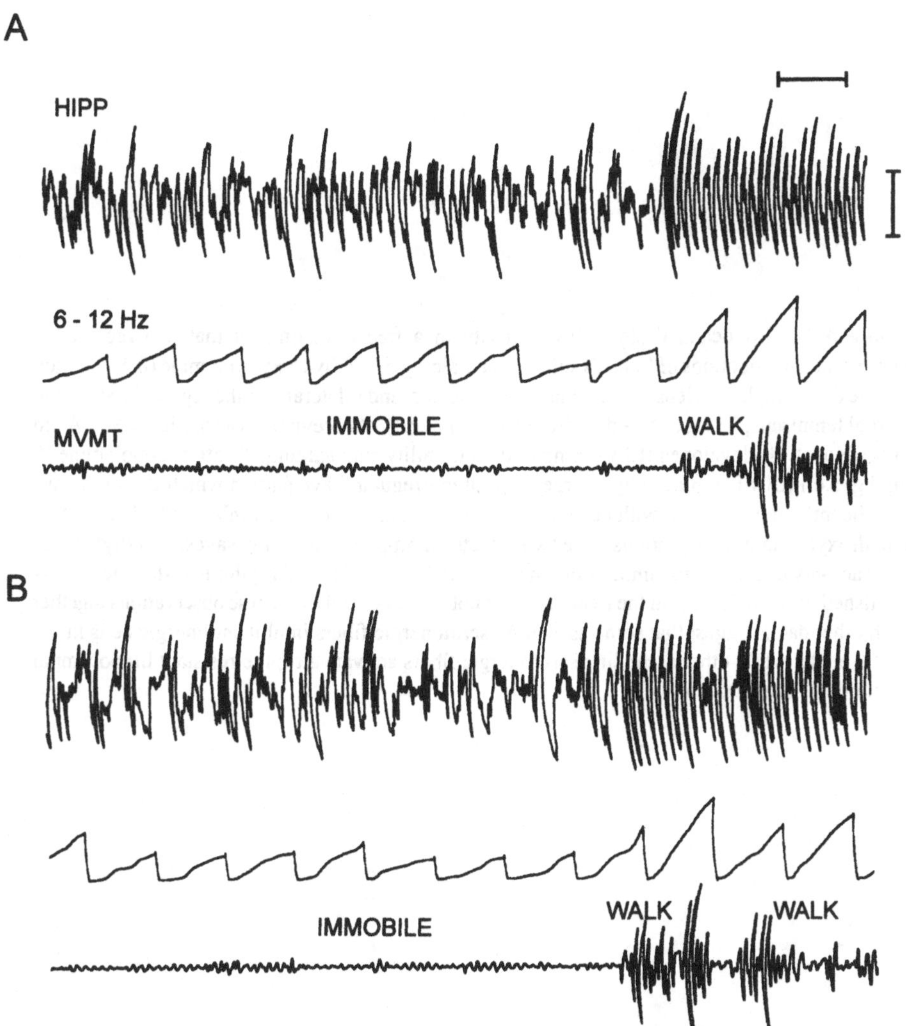
A
HIPP
6 - 12 Hz
MVMT
IMMOBILE
WALK
B
IMMOBILE
WALK
WALK

Figure 6-2. Hippocampal slow wave activity in a freely moving rat that had received 4 intrabrainstem injections of Locke's solution containing 5,7- dihydroxytryptamine (0.5μl in each of: the dorsal raphe nucleus, the median raphe nucleus, and bilaterally in the region dorsal to the medial lemniscus; 25 μg of 5,7- dihydroxytryptamine/μl). Abbreviations as in Figure 6-1. A: no drug. Note well-developed RSA during both immobility and walking. B: after scopolamine (5 mg/kg, s.c.) RSA is replaced by a large amplitude irregular wave pattern which does not vary significantly in correlation with behavior. Rat 2282. Calibration: 1s, 1 mV. Note that the 5,7-dihydroxytryptamine injections have two effects on hippocampal slow waves: (1) rhythmical slow activity occurs during immobility in the treated rats; and (2) all rhythmical slow activity is abolished by scopolamine in the treated rat but not in the control rat. These observations together with other data, suggest that some ascending serotonergic fibers inhibit cholinergic cells in the septal nuclei while other ascending serotonergic fibers activate atropine-resistant hippocampal rhythmical slow activity.

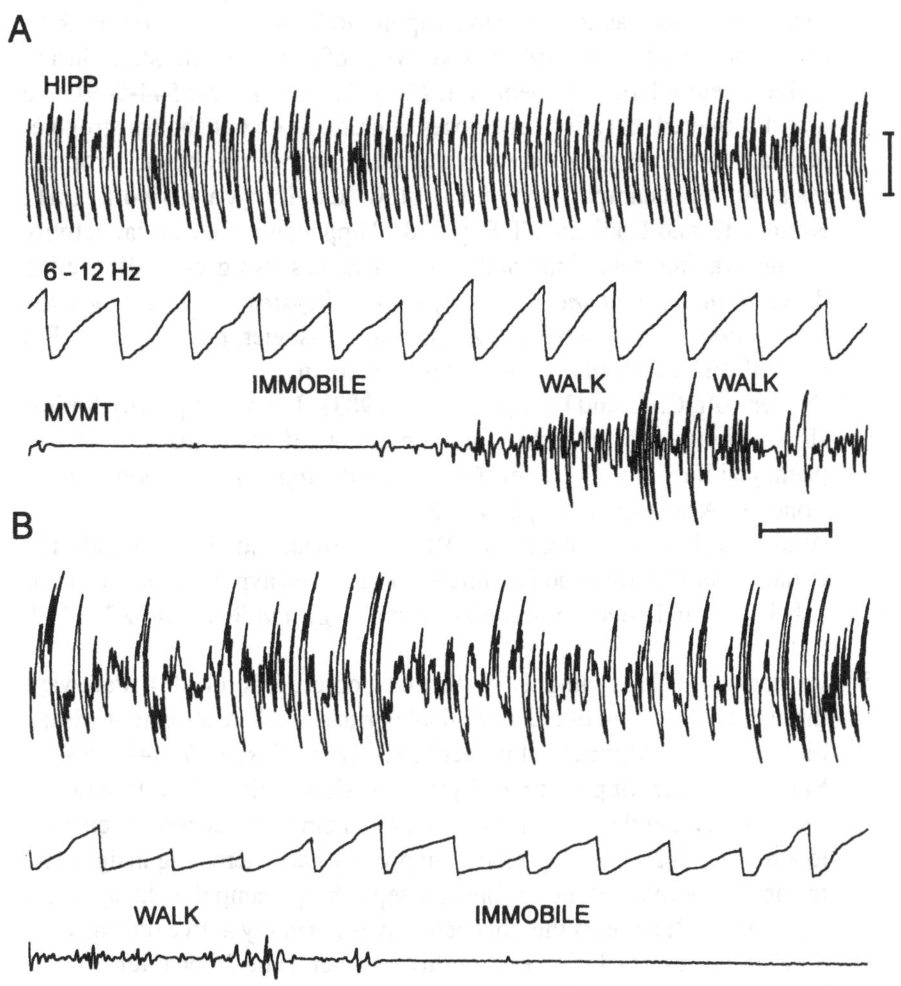
A
HIPP
6 - 12 Hz
IMMOBILE
WALK
WALK
MVMT
B
WALK
IMMOBILE

neocortical electrical activity, and relations to behavior. *Experimental Neurology, 61*: 485-515.

2. During this period several other notable discoveries were made. Terry showed that the higher frequency hippocampal rhythmical slow activity associated with the muscular twitches of active sleep was resistant to atropine while the lower frequency activity appearing during the inter-twitch intervals was sensitive to atropine [Robinson, T.E., Kramis, R.C. and Vanderwolf, C.H. (1977). Two types of cerebral activation during active sleep: relations to behavior. *Brain Research, 124*: 544-549]. We also discovered that anesthetic doses of urethane do not abolish atropine-resistant low voltage fast activity even though they do abolish atropine-resistant hippocampal rhythmical slow activity [Vanderwolf, C.H., Kramis, R. and Robinson, T.E. (1978). Hippocampal electrical activity during waking behaviour and sleep: analyses using centrally acting drugs. *Functions of the septo-hippocampal system, Ciba Foundation Symposium. #58(new series)*, Amsterdam; Elsevier, pp. 199-226]. The basis of this selectivity is completely unknown.
3. Vanderwolf, C.H., and Leung, L-W.S. (1983). Hippocampal rhythmical slow activity: a brief history and the effects of entorhinal lesions and phencyclidine. In W. Seifert (ed.) *Neurobiology of the hippocampus*, London: Academic Press, 275-302.
4. Whishaw, I.Q. and Kolb, B. (1979). Neocortical and hippocampal EEG in rats during lateral hypothalamic lesion-induced hyperkinesia: relations to behavior and effects of atropine. *Physiology and Behavior, 22*: 1107-1113.
5. Vanderwolf, C.H., Leung, L.-W.S., and Cooley, R.K. (1985). Pathways through cingulate, neo-and entorhinal cortices mediate atropine-resistant hippocampal rhythmical slow activity. *Brain Research, 347*: 58-73. Since all of the hippocampal rhythmical slow activity in rats with the long coronal cut through the neocortex and cingulate cortex is sensitive to atropine, this surgical preparation provides an interesting insight into the normal activity of the cholinergic septo-hippocampal pathway acting in isolation. It appears that this pathway is normally active during Type 1 behavior but not during immobility or other Type 2 behavior.
6. Buzsaki, G., Leung, L.W.S., and Vanderwolf, C.H. (1983). Cellular bases of hippocampal EEG in the behaving rat. *Brain Research Reviews, 6*: 139-171.
7. Buzsaki, G., and Vanderwolf, C.H. (eds) (1985). *Electrical activity of the archicortex*. Budapest: Akadémiai Kiadó.
8. Vanderwolf, C.H., and Baker, G.B. (1986). Evidence that serotonin mediates non-cholinergic neocortical low voltage fast activity, non-

cholinergic hippocampal rhythmical slow activity and contributes to intelligent behavior. *Brain Research*, *374*: 342-356

Chapter 7

Ascending Cholinergic Control of the Neocortex and Hippocampus

In 1967, two British neuroscientists, P.R. Lewis and C.C.D. Shute, published a pair of papers[1] on what they called the "ascending cholinergic reticular system", a system of ascending cholinergic projections from the brainstem to the cerebral cortex. It was known at the time that drugs (anticholinesterases) that inhibit acetylcholinesterase, the enzyme involved in the breakdown of acetylcholine, produce low voltage fast activity in the neocortex and rhythmical slow activity in the hippocampus (neocortical and hippocampal activation). Conversely drugs that block muscarinic cholinergic transmission in the brain block at least one component of neocortical and hippocampal activation (see Chapters 4, 5 and 6). Therefore, it seemed reasonable to think that the actions of anticholinesterases and anti-muscarinic drugs on neocortical and hippocampal activity were due to interference with the normal function of an ascending cholinergic activating system.

The discovery of atropine-sensitive and atropine-resistant components of neocortical and hippocampal activation suggested that it might be possible to block the cholinergic component of activation by appropriately selective surgical procedures. This was important because atropine and other antimuscarinic drugs could be acting at many different points in the brain when they are injected systemically. A more selective procedure was needed to discover the precise site of action. However, the anatomical papers of Shute and Lewis did not suggest any obvious point of attack. Part of the difficulty arose from the fact that Shute and Lewis had based their conclusions on data obtained by the use of a staining procedure that visualised the location of acetylcholinesterase. Some years later when it also become possible to localise choline acetyltransferase, the enzyme involved in the synthesis of acetylcholine, it became apparent that some brain neurons that are clearly not cholinergic because they do not synthesize (and therefore, presumably, cannot release) acetylcholine, nonetheless contain large amounts of acetylcholinesterase. Therefore, the Shute and Lewis "cholinergic reticular system" undoubtedly contained many neurons that were not part of a cholinergic activating input to the cerebral cortex.

The situation was greatly clarified by the discovery of a population of neurons in the basal forebrain region (substantia innominata and nucleus basalis of Meynert) that contained both choline acetyltransferase and acetylcholinesterase and sent axons to virtually the entire neocortex and cingulate cortex. These neurons attracted the interest of the neuroscientific community around 1980 when it became apparent that they were destroyed with some degree of selectivity in cases of Alzheimer's disease[2].

In the fall of 1981, Dwight Stewart, a student from my alma mater, the University of Alberta in Edmonton, arrived to begin work on a master's thesis. The functions of the ascending cholinergic projections from the basal forebrain provided a suitable topic. Dwight, Richard Cooley and I began by taking records of neocortical activity in rats with unilateral surgical destruction of the basal forebrain area. There were two good reasons for making these lesions on one side of the brain only. (1) According to neuroanatomical studies, the projections to the neocortex are entirely unilateral, making it possible to affect activity in one hemisphere while the other hemisphere serves as a control. (2) Bilateral destruction of an area just caudal to the basal forebrain tends to result in a potentially lethal pulmonary edema. We hoped to avoid any chance of this.

Surgical destruction of the basal forebrain produced only a severe depression of electrocortical activity in the ipsilateral hemisphere, an effect I had no wish to pursue at that time[3]. More interesting results were obtained by local injections of kainic acid, a neurotoxin that destroys most neurons but has little direct effect on axons. Therefore, we hoped, kainic acid would destroy basal forebrain cholinergic neurons without producing extensive damage to the many fibers of passage that traverse this region of the brain.

The experiment worked beautifully the first time we tried it[4]. After recovery from the immediate post-operative effects, the neocortex ipsilateral to the kainic acid injection displayed an abundance of large irregular slow waves in the waking state much like those produced by injections of atropine, while the contralateral (control) hemisphere displayed mainly low voltage fast activity (Figure 7-1). Further, like the effects of atropine, the kainic acid-induced slow waves were much more prominent during immobility than during head

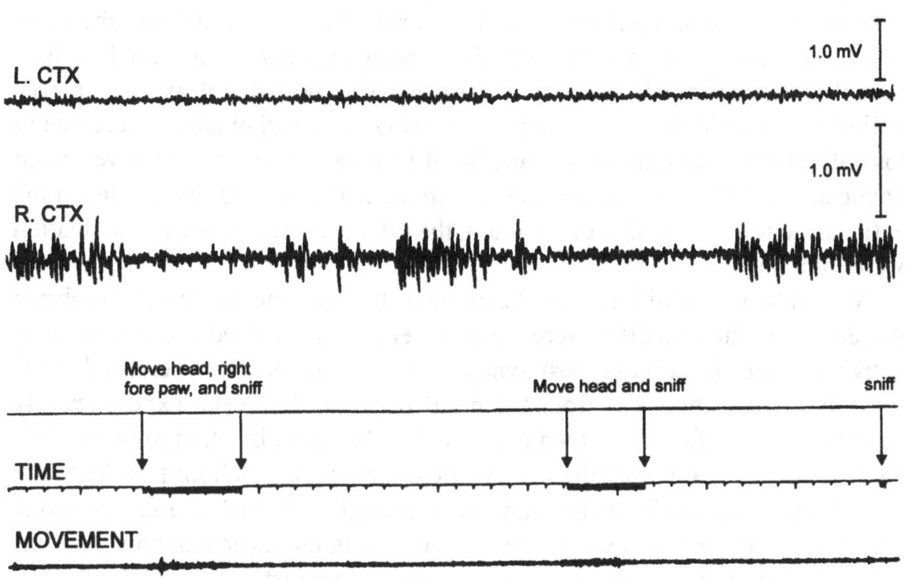

Figure 7-1. Records from surface-to-depth electrode pairs in left (L. CTX) and right (R. CTX) parietal neocortex in a rat taken 25 days after injection of kainic acid into the right basal forebrain. This results in a loss of cholinergic projections to the ipsilateral cerebral cortex. Time in seconds. MOVEMENT refers to the output of a magnet-and-coil type of movement sensor. Note that large amplitude slow waves occur in the right cortex during immobility and sniffing but that low voltage fast activity occurs during head and forelimb movements accompanied by sniffing. Low voltage fast activity is continuously present in the left neocortex. From Vanderwolf, C.H. (1988). Cerebral activity and behavior: control by central cholinergic and serotonergic systems. *International Review of Neurobiology, 30*: 255-340

movements or locomotion. Post-mortem histological studies showed a marked loss of cells in the basal forebrain and fibers in the neocortex that took up the stain for acetylcholinesterase. These experiments suggested that the ability of a normal rat to maintain its neocortex in an activated or low voltage fast condition during behavioral immobility is dependent on the ascending cholinergic projections from the basal forebrain to the neocortex.

Cells in many thalamic nuclei are exquisitely sensitive to kainic acid and disappeared in large numbers when the neurotoxin was injected into the basal forebrain. Since the thalamus had long been regarded as a vital link in a reticulothalamocortical pathway mediating neocortical activation, it was definitely possible that the unilateral slow waves we had obtained were due to loss of thalamic neurons rather than basal forebrain neurons. However when kainic acid injections were made locally in the thalamus, destroying almost the entire structure on one side of the brain, the effects on electrocortical activation were minimal.

We obtained additional evidence that the cholinergic basal forebrain projections to the neocortex were the ones genuinely involved in electrocortical activity. It was known that post-synaptic muscarinic receptors persisted in the neocortex when the input from the basal forebrain had been experimentally destroyed. Therefore, it seemed possible that the administration of muscarinic agonists, drugs that mimic the muscarinic effects of acetylcholine in the brain, would restore neocortical activation after the basal forebrain had been destroyed. Pilocarpine, a classical example of a direct-acting muscarinic agonist had this effect, as did oxotremorine, a second muscarinic agonist[4,5]

Further, Allan Fine at Dalhousie University in Nova Scotia, suggested an experiment in which basal forebrain cells from rat embryos were injected in the region adjacent to the recording electrodes in rats that had previously received kainic acid injections into the basal forebrain. We hoped that the kainic acid would eliminate low voltage fast activity in the neocortex during behavioral immobility (as already shown) and that the cholinergic cells of the basal forebrain would subsequently restore such activity when they were injected next to the recording electrodes. Control rats received either no injection or an injection of hippocampal cells. The experiment appeared to work. The neocortex sites injected with embryonic basal forebrain tissue showed a good ingrowth of acetylcholinesterase-positive fibers and more low voltage fast activity than the sites that received no embryonic cells or embryonic cells derived from the hippocampus[6]. This suggests that local release of acetylcholine in the neocortex can produce low voltage fast activity. Of course it is true that the implants of basal forebrain tissue may release transmitters other than acetylcholine, but other evidence also indicates the essential involvement of acetylcholine release. As John Szerb of Dalhousie University was the first to show, more acetylcholine is released in the neocortex during periods of low voltage fast activity than during periods of large amplitude slow wave activity. Further, it is known that application of atropine to the surface of the neocortex can induce local large amplitude slow wave activity while the similar application of the anticholinesterase drug eserine or acetylcholine itself can produce low voltage fast activity. This indicates that acetylcholine produces neocortical activation by a direct local action in the neocortex.

All these data suggest strongly that the cholinergic projections from the basal forebrain to the neocortex are responsible for the generation of atropine-sensitive low voltage fast activity. More definitive evidence could be obtained if it were possible to study the relation of the action potentials of basal forebrain cholinergic neurons to the occurrence of spontaneous or elicited neocortical low voltage fast activity. Although there was no simple method available to identify the neurochemical nature of a recorded basal forebrain neuron, it seemed possible to take advantage of the fact that most, perhaps all, of the neurons in the rat's basal forebrain that send axons to the neocortex also contain the enzyme choline acetyltransferase. This means that a brief pulse of electrical stimulation applied to the cortex should fire a basal forebrain cholinergic neuron antidromically (i.e. contrary to the natural direction of action potential propagation) but that non-cholinergic neurons should not be fired antidromically.

To pursue this project, I invited Laszlo Detari of Eötvös Loránd University in Budapest to spend a year with us at Western. The project proved to be a great success: basal forebrain neurons that could be fired antidromically by stimulation of the neocortex were much more active during periods of neocortical low voltage fast activity than during periods of large amplitude slow waves (Figure 7-2). Some of the cells that could not be fired antidromically (perhaps they were local interneurons or sent axons to other structures) were more active during periods in which large slow waves occurred in the neocortex than during periods in which low voltage fast activity was present. This provided strong evidence that basal forebrain cholinergic neurons were responsible for atropine-sensitive neocortical activation.

While these experiments on the basal forebrain cells and neocortical activity were in progress, Dwight Stewart began a series of experiments on the effect of ibotenic acid injections in the septal nuclei on hippocampal activity in chronically prepared rats. We hoped that ibotenic acid would destroy cholinergic septo-hippocampal projection cells while sparing fibers of passage and that this effect would abolish hippocampal rhythmical slow activity in urethane anesthetized rats without abolishing hippocampal rhythmical slow activity in waking rats during the performance of Type 1 movement. At first the experiment seemed to work: rhythmical slow activity during urethane anesthesia was indeed abolished while rhythmical slow activity during locomotion in the

Figure 7-2. An example of activity in an antidromically identified cortically projecting cell. A: collision test. Full response at a delay of 11 ms, but no antidromic invasion at 9 ms. Ten successive stimuli. Calibration, 10 ms. B: electrode track and cell localization in the camera lucida drawing of the histological section. C: Polygraph record of neocortical slow wave activity (contaminated by hippocampal theta waves) together with pulses triggered by unitary action potentials. The gradual appearance of large slow waves in the EEG is paralleled by a marked decrease in cell firing. D: after approximately 70s of continuous slow wave activity, during which the cell almost stopped firing, a tail pinch elicits cortical activation and high frequency firing in the cell. Calibration, 0.25 mV and 5s. From Detari, L., and Vanderwolf, C.H. (1987). Activity of identified cortically projecting and other basal forebrain neurons during large slow waves and cortical activation in anaesthetized rats. *Brain Research, 437*: 1-8 with the permission of Elsevier Science.

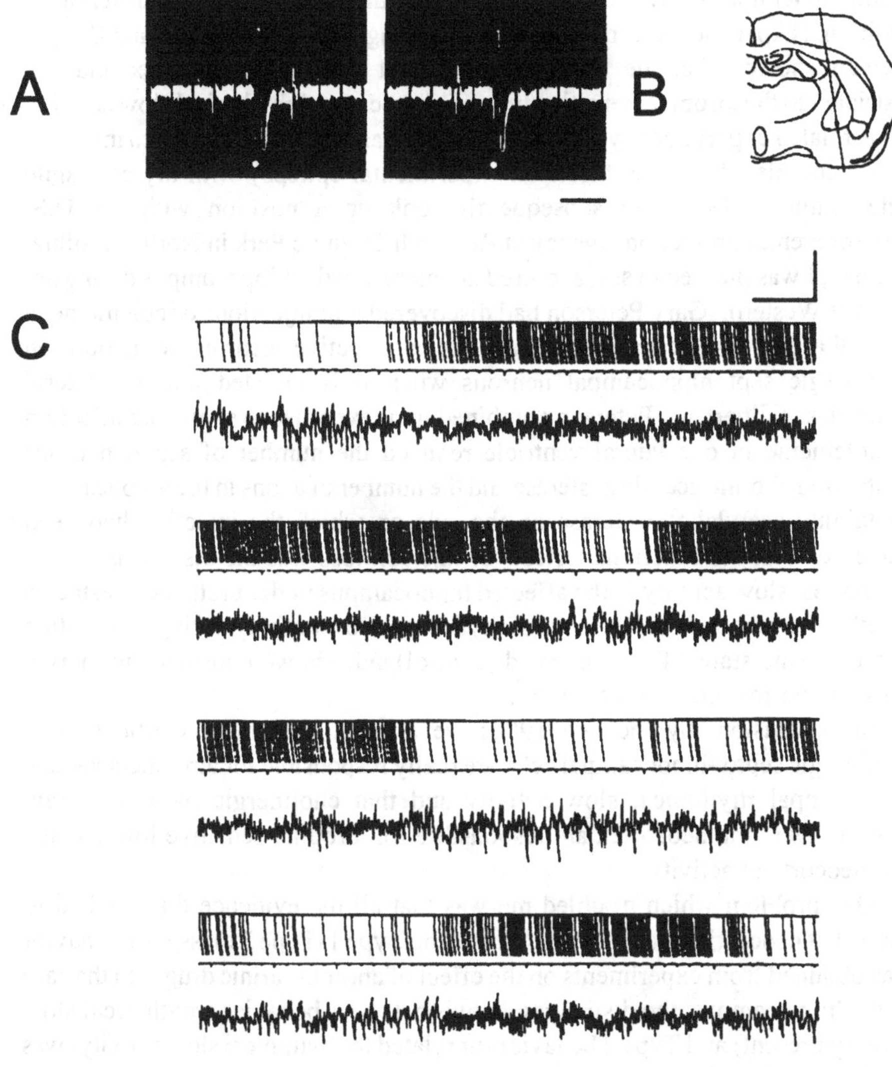
A
B
C

waking state persisted. However, subsequent histological study showed that acetylcholinesterase containing cells in the septal nuclei (the medial septal nucleus and the nuclei of the diagonal band of Broca) were not much affected by the ibotenic acid. Instead there was severe cell loss in a dorsomedial part of the septal nuclei that receives a strong input *from* the hippocampus. Furthermore, additional experiments carried out by Stan Leung, Lee-Anne Martin, and Dwight Stewart showed that the ibotenic acid effect was rather complex and not restricted to the atropine-sensitive type of hippocampal rhythmical slow activity.

A related experiment by Mary Gilbert and Gary Peterson is easier to interpret. Mary completed a Ph.D. thesis on experimental epilepsy with my colleague Peter Cain in 1984 and subsequently took up a position with the U.S. Environmental Protection Agency at Research Triangle Park in North Carolina. She had, I was pleased to see, acquired an interest in the hippocampus during her years at Western. Gary Peterson had discovered that injections of colchicine, a drug that blocks axonal transport, has a selective neurotoxic action on cholinergic septohippocampal neurons when it is injected into the lateral ventricles. Gilbert and Peterson combined their expertise to show that injection of colchicine in one lateral ventricle reduced the number of septal neurons containing choline acetyltransferase and the number of axons in the hippocampus containing acetylcholinesterase on the side on which the injection had been made. One of the functional effects of this cell loss was the disappearance of rhythmical slow activity in the affected hippocampus under urethane anesthesia together with a good preservation of rhythmical slow activity during locomotion in the waking state. The uninjected (control) side showed normal cholinergic cells and normal electrical activity[7].

In conclusion, by the late 1980s we all felt reasonably confident that cholinergic septo-hippocampal cells are really responsible for atropine-sensitive hippocampal rhythmical slow activity and that cholinergic basal forebrain projections to the neocortex are responsible for atropine-sensitive low voltage fast neocortical activity.

One problem which troubled me was that all the evidence that we had to show that neocortical activity is related to the Type 1-Type 2 classes of behavior was obtained from experiments on the effect of antimuscarinic drugs. In the case of the hippocampus, the distinction between Type 1 behavior (rhythmical slow activity present) and Type 2 behavior (unrelated to rhythmical slow activity) was obvious in an undrugged normal rat but in the neocortex the low voltage fast activity during Type 2 behavior was usually very similar to the low voltage fast activity occurring during Type 1 behavior. The conclusion that neocortical activity was really organized in relation to the Type1-Type 2 behavior distinction would be considerably strengthened if it could be demonstrated in some way in undrugged rats.

There were hints of various kinds. Jerry Schwartzbaum and his colleagues at the University of Rochester in New York had shown that a late component of the striate cortex response to a light flash is reduced in amplitude during the presence of rhythmical slow activity in the hippocampus in freely moving rats. L. Pickenhain and F. Klingberg of Karl-Marx University in East Germany had shown differences in the striate cortex response to a light flash depending on whether the rat was engaged in "comfort movements" (grooming, presumably) or "intended movements" (perhaps this referred to walking or other Type 1 movements).

However, there are many complications involved in such experiments. The amplitude of the cortical response to a flash of light depends on many factors, such as the direction the rat is looking when the flash occurs, the sensitivity of the retina (light adapted or dark adapted) whether the rat happens to be blinking or not, the diameter of the pupil, and possible variations in transmission through subcortical synapses in the retina and the lateral geniculate body.

A superior method of studying the relation between neocortical activity and behavior was provided by the transcallosal evoked potential. It had been known for some time that a brief single pulse of electric current applied to one side of the neocortex (either left or right) in a waking animal produces a complex response in the mirror-image point in the opposite side of the neocortex. We studied this effect in both frontal (motor) cortex and in parietal (somatosensory) cortex, obtaining very similar results in both regions. The early component of the transcallosal response consists of excitatory post-synaptic potentials and unit discharges while a later component consists of inhibitory post-synaptic potentials associated with a widespread suppression of unit discharges for a period of about 100 milliseconds. An incidental observation made by Ron Racine and his colleagues at McMaster University in Hamilton, Ontario, was that this late potential varied in a remarkable way depending on the behavior in progress at the time. Stan Leung, Greg Harvey (a graduate student working on a master's degree) and I decided to study this effect in detail using a system in which the behaving rat and the neocortical response were recorded simultaneously on videotape[8]. Although the amplitude of both the early and late components of the transcallosal response varied in relation to the behavior in progress when the evoked potential was recorded, the changes in the late component were especially striking (Figures 7-3, 7-4). During waking immobility a large amplitude long duration potential occurred in correlation with a suppression of multiunit activity. Active Type 2 behaviors such as face-washing, drinking, or biting and chewing food were associated with a slightly reduced late component but during Type 1 behavior such as head movements, gross changes in posture, or walking, the late component was reduced by 50-100 percent of what it was during waking immobility. In association with this, the period of multiunit

Figure 7-3. Effect of motor activity on the transcallosal response. EP, bipolar transcortical record of slow wave evoked response in left sensorimotor cortex, deep positivity up; MU, multiunit activity derived from the deep member of the same bipolar electrode pair; S, single pulses of monopolar stimulation of right sensorimotor cortex at 2 x threshold (228 μA). *Top*: waking immobility. *Bottom*: elicited walking. Note the large late deep-positive component of the evoked potential and associated multiunit suppression during immobility and small late deep-positive component and associated multiunit activity during elicited walking. The early deep-negative component is only slightly altered. Four superimposed single sweeps in each behavioral condition. Calibration: EP, 0.5 mV; MU, 0.05 mV; 20 ms. Figures 7-3 and 7-4 are from Vanderwolf, C.H., Harvey, G.C., and Leung, L.-W.S. (1987). Transcallosal evoked potentials in relation to behavior in the rat: effects of atropine, *p*-chlorophenylalanine, reserpine, scopolamine, and trifluoperazine. *Behavioural Brain Research, 25*: 31-48, with the permission of Elsevier Science.

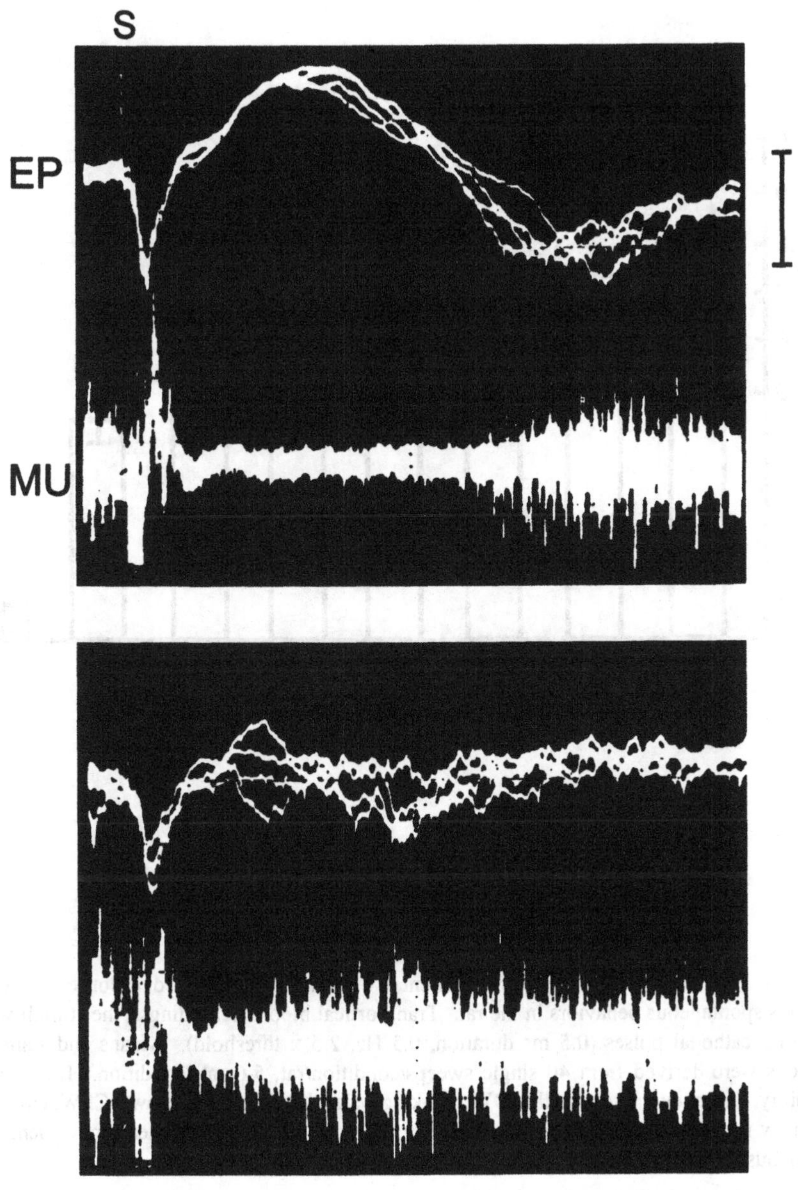
S
EP
MU

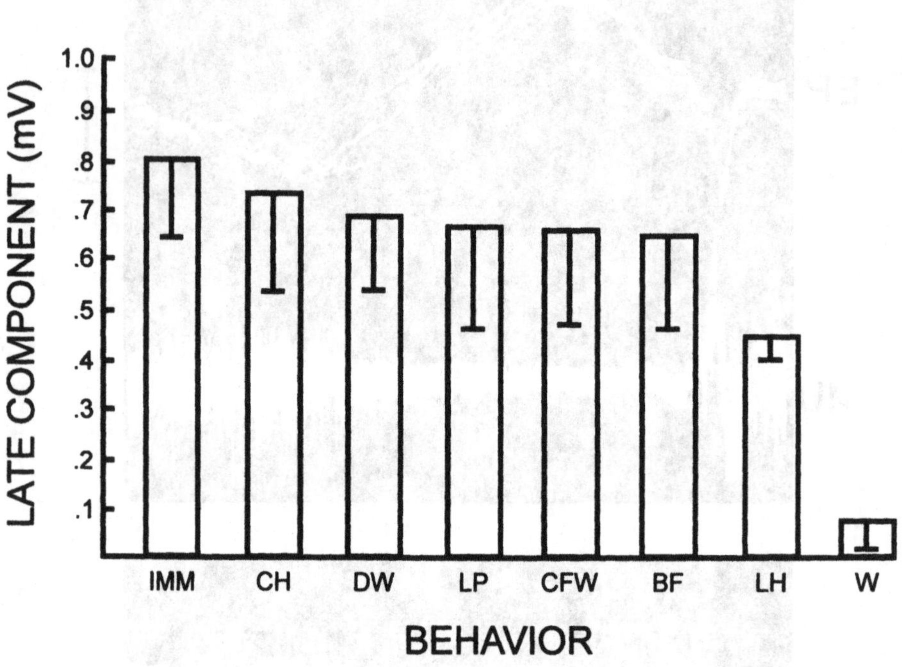

Figure 7-4. Peak amplitude of the late component of the transcallosal evoked response in relation to various spontaneous behaviors in the rat. Transcortical bipolar recordings: the stimuli were monopolar cathodal pulses (0.5 ms duration, 0.5 Hz, 2.5 x threshold). Means and standard deviations were derived from 40 single sweeps/condition/rat, 5-6 rats/condition. IMM, alert immobility; CH, chewing food pellet; DW, drinking water; LP, licking forepaws; CFW, circus or rotary paw movements during face-washing; BF, biting food; LH, lateral head movement; W, spontaneous walking.

suppression that was ordinarily present during Type 2 behavior was reduced or completely eliminated. Since these effects could not be duplicated by purely passive movements, they are unlikely to be due to sensory feedback from movement. Thus, the data showed that a neocortical inhibitory response that is ordinarily present during Type 2 behavior is reduced or eliminated during Type 1 behavior. Therefore in the intact undrugged brain, the activity of the neocortex, as well as the hippocampus, is closely related to the occurrence of Type 1 behavior.

The cingulate cortex, as well as the hippocampus and neocortex, is active in close relation to Type 1 behavior. Stan Leung and J.G.G. Borst, a post-doctoral fellow from the Netherlands who worked with Stan for a few months, showed that the cingulate cortex generates 6-10 Hz rhythmical waves during walking, rearing, postural shifts, head movements and rapid-eye-movement sleep while an irregular wave pattern containing 20 ms duration sharp waves is present during grooming, drinking, eating, alert immobility and quiet sleep.[9] Therefore, it is probable that the entire dorsal cortical mantle is active in relation to Type 1 behavior as opposed to Type 2 behavior.

Stan, Greg Harvey and I carried out pharmacological studies on the transcallosal evoked potential and these were later extended by Hans Dringenberg, a graduate student who arrived from the University of Lethbridge in 1991. The results supported the hypothesis that both cholinergic muscarinic and serotonergic inputs to the neocortex controlled the behavior-related variation in the transcallosal evoked potential.

Dwight Stewart successfully defended his Ph.D. thesis in 1986 and returned to Edmonton to spend a year working in neurochemistry with Glen Baker at the University of Alberta. During this year, Dwight became very discouraged by the employment prospects in neuroscience. It seemed to be all long hours, low pay, and little prospect of an attractive position anywhere. Consequently, he enrolled in medical school at the University of Calgary, and emerged a few years later as a neurologist with much better pay and many attractive employment options.

Notes on Chapter 7

1. Lewis, P.R., and Shute, C.C.D. (1967). The cholinergic limbic system: Projection to hippocampal formation, medial cortex, nuclei of the ascending cholinergic reticular system, and the subfornical organ and supraoptic crest. *Brain, 90*: 521-540.
   Shute, C.C.D., and Lewis, P.R. (1967). The ascending cholinergic reticular system: neocortical, olfactory and subcortical projections. *Brain, 90*: 497-520.
2. Davies, P., and Maloney, A.J.F. (1976). Selective loss of central cholinergic neurons in Alzheimer's disease. *Lancet*, 1403.

Whitehouse, P.J., Price, D.L., Clark, A.W., Coyle, J.T., and DeLong, M.R. (1981). Alzheimer's disease: evidence for selective loss of cholinergic neurons in the nucleus basalis. *Annals of Neurology, 10*: 122-126.

3. This effect is reminiscent of the severe depression of slow wave activity (both the irregular waves and the rhythmical waves) in the hippocampus when the septal nuclei are destroyed. The substantia innominata and nucleus basalis of Meynert appear to stand in the same relation to the neocortex as the septal nuclei do to the hippocampus.

4. Stewart, D.J., MacFabe, D.F., and Vanderwolf, C.H. (1984). Cholinergic activation of the electrocorticogram: Role of the substantia innominata and effects of atropine and quinuclidinyl benzilate. *Brain Research, 322*: 219-232.
5. Vanderwolf, C.H., Raithby, A., Snider, M., Cristi, C., and Tanner, C. (1993). Effects of some cholinergic agonists on neocortical slow wave activity in rats with basal forebrain lesions. *Brain Research Bulletin, 31*: 515-521.
6. Vanderwolf, C.H., Fine, A., and Cooley, R.K. (1990). Intracortical grafts of embryonic basal forebrain tissue restore low voltage fast activity in rats with basal forebrain lesions. *Experimental Brain Research, 81*: 426-432.
7. Gilbert, M.E., and Peterson, G.M. (1991). Colchicine-induced deafferentation of the hippocampus selectively disrupts cholinergic rhythmical slow wave activity. *Brain Research, 564*: 117-126.
8. Vanderwolf, C.H., Harvey, G.C., and Leung, L.-W.S. (1987). Transcallosal evoked potentials in relation to behavior in the rat: effects of atropine, *p*-chlorophenylalanine, reserpine, scopolamine, and trifluoperazine. *Behavioural Brain Research, 25*; 31-48.
9. Leung, L.-W.S., and Borst, J.G.G. (1987). Electrical activity of the cingulate cortex. I. Generating mechanisms and relations to behavior. *Brain Research, 407*: 68-80.
Borst, J.G.G., Leung, L.-W.S., and MacFabe, D.F. (1987). Electrical activity of the cingulate cortex. II. Cholinergic modulation. *Brain Research, 407*: 81-93.

Chapter 8

What is the Function of Cerebral Cortical Activation?

The preceding chapters have shown that many aspects of cerebral cortical electrical activity occur in close temporal correlation with the details of motor activity. However, the demonstration of a correlation between a brain electrical event and the occurrence of a movement is not sufficient to establish the significance of the cerebral electrical event in question. The fact that two events are correlated does not necessarily mean that one is the cause of the other. Both may be caused by some unknown third factor. This logical difficulty can be surmounted if the hypothesized causal factor can be eliminated in some way. For example, does surgical or pharmacological blockade of cerebral electrocortical activation have an effect on behavior? If it does, such activation may play a role in causing the behavior.

Classical studies of the functional significance of electrocortical activation showed that large surgical lesions of the brainstem reticular formation produced a somnolent or comatose condition together with a reduction, but not a total elimination, of the occurrence of low voltage fast activity in the neocortex[1]. This was taken to mean that elimination of cortical activation resulted in coma or sleep. However, surgical lesions of the brainstem cannot be regarded as selective for the ascending pathways involved in cortical activation: inevitably a great many unrelated pathways will be destroyed as well. These pathways may have been responsible for the behavioral effects observed. The discovery that a combination of central muscarinic and central serotonergic blockade can completely eliminate both neocortical low voltage fast activity and hippocampal rhythmical slow activity permits a re-evaluation of the significance of cerebral cortical activation patterns uncomplicated by the effects of extensive nonspecific destruction of brainstem neurons.

This topic can be understood only in terms of its relation to the general question of how the central nervous system generates behavior. As an example of this, consider the behavior of a rat which has not eaten for some time. Acting under the influence of a very complex set of factors including sensory inputs from viscera and from the environment, direct hormonal influences, and, perhaps, blood levels of glucose and other nutrients, the brain must generate: (1) motor responses that bring the rat near enough to food to permit mouth contact; and (2) biting, chewing and swallowing. As C.S. Sherrington[2] pointed out many years ago, behavioral responses of the first group are generally under the control of distance receptors such as vision or olfaction while those in the second group are generally under the control of contact receptors. Section of the sensory trigeminal fibers that innervate the lips and snout prevent the jaw opening and biting reflexes normally elicited by contact with these structures. Feeding is severely impaired. Rats that have undergone section of sensory nerves to the lips and snout may launch repeated predatory attacks on a mouse without inflicting any harm beyond covering the smaller animal with saliva[3]. In this case, predatory locomotor activity persists and is still under good stimulus control (the rat walks to the mouse) but the reflexive biting behavior is suppressed.

The behaviors involved in bringing the reflexogenic areas of the lips and snout in contact with food vary greatly, depending on the circumstances, and include walking, running, jumping, climbing, swimming, digging, and manipulation of objects with the forelimbs (e.g. pressing a lever in a Skinner box). These are all Type 1 behaviors as discussed in Chapter 3. Under some circumstances these Type 1 behaviors may be involved in setting the stage for the activation of reflexes subserving eating or drinking; under other circumstances they may play a role in creating a situation in which copulatory reflexes are elicited, one in which nesting material is nibbled into shreds, maternal care is provided, or a predator is avoided. Presumably, the selection of specific behaviors and their performance in a manner which is appropriate to particular circumstances is managed by complex neural circuitry that has developed under the influence of the genome (permitting untrained instinctive behavior) but is also capable of modification by experience (permitting learned behavior). It seems likely that the neural regulation of Type 1 behavior is far more complex and more heavily dependent on the cerebrum than the more reflexive activities included in Type 2 behavior.

A fundamental insight into the way the nervous system is organized in relation to behavior has been provided by studies in which the capabilities of different parts of the brain or spinal cord have been studied in isolation. Studies by C.S. Sherrington[2], many years ago, showed that the isolated spinal cord mediates a wide variety of reflex responses that are easily recognisable as components of various patterns of normal behavior. For example a spinal dog (in which connections between the brain and the spinal cord have been severed) will exhibit rhythmical stepping movements of the hind limbs, obviously a component of normal locomotion, if appropriate sensory input is provided. Further, a light punctate moving contact on the thoracic skin of a spinal dog (mimicking the effect of a crawling louse) elicits rhythmical scratching movements of the ipsilateral hind leg. This is clearly a component of normal self-grooming behavior.

More complex patterns of behavior are observed in high decerebrate animals in which the forebrain is disconnected or removed, but the brain stem, cerebellum and spinal cord are still able to control motor activity[4]. Unlike spinal animals, a high decerebrate rat is able to right itself from a reclining position and walk about easily on a flat surface. It can rear up on its hind legs, sit up and wash its face, and will lick or bite in response to appropriate stimulation of its lips and snout. Despite this, high decerebrate rats make no attempt to feed themselves. Decorticate rats, in which the neocortex, cingulate cortex and hippocampal formation have been surgically removed, are rather similar to high decerebrate rats with respect to behavior although the deficits are less severe[5]. What seems to be lacking in these preparations with respect to feeding, for example, is a mechanism which, in normal rats, ensures that Type 1 behavior such as locomotion and head movement, guided by visual or olfactory stimuli, brings the reflexogenic zones of the snout and lips into contact with food so that the more reflexive responses of biting, chewing and swallowing can occur. In the same way, special postures of the trunk and head must be assumed and maintained in order to permit reflexive licking and biting of the fur during grooming of the body; locomotion, head movement, mounting, etc., occurring under stimulus control, are required to bring the genitalia of a male rat into contact with the genitalia of a receptive female rat in order to activate copulatory reflexes. The adaptive performance of all these Type 1 behaviors is severely impaired in animals following extensive injury to the cerebrum.

The conclusion suggested by all these observations is that behavior control mechanisms are hierarchically organized. Relatively simple reflexes, such as those exhibited by the isolated spinal cord, are organized into larger patterns such as walking, rearing, or face-washing by circuits located primarily in the cerebellum and brain stem. These brain stem and cerebellar patterns, in turn, are controlled by circuits in the forebrain which generate Type 1 behavior patterns appropriate to current environmental conditions and the hormonal and nutritional status of the animal. The Type 1 behavior patterns then help establish conditions under which Type 2 reflexive or consummatory sensori-motor reactions can occur.

There are many pathways by which the neocortex can control the subcortical circuits involved in locomotion, shifts in posture, and other gross motor activities. This includes projections from virtually the entire neocortex to the pons and the striatum as well as projections to the thalamus and a variety of brain stem targets. It has been known for many years that visual cortical control of motor activity is exerted primarily via descending projections to brainstem structures.[6] The hippocampal formation too has massive descending projections to regions of the hypothalamus and brain stem which are involved in gross motor activity.[7]

I had assumed for a long time that if the activation patterns of the hippocampal formation (rhythmical slow waves) and of the neocortex (low voltage fast activity) could be selectively blocked, then the normal function of these cortical structures would be grossly impaired, producing a behavioral syndrome similar to the one produced by decortication or even high decerebration. In fact, one reason for pursuing the question of the neurochemical identity of the ascending activating systems so persistently was that I hoped to demonstrate the true behavioral effect of a selective blockade of ascending activation.

By the early 1980s I had begun to despair of ever discovering the real identity of the ascending activating systems. Nonetheless, wishing to have at least some behavioral data relevant to the function of these ascending activating systems I decided to accept a second-choice solution. It had been known since 1964 that reserpine will totally abolish the behavior of lever pressing reinforced by electrical stimulation of the lateral hypothalamus. However, a subsequent injection of *d*-amphetamine will temporarily restore a normal rate of lever pressing. The effect is truly impressive. Over the course of a few minutes, an immobile reserpinized rat in which lateral hypothalamic stimulation has very little effect, is transformed by amphetamine into a rat which will walk toward the lever, making appropriate orienting movements of the head, and begin to press the lever with its forepaws or with mouth and forepaws together, just as in a normal undrugged rat. The basis of the restoration of both spontaneous motor activity and directed lever-pressing behavior seems to be that both are dependent on the release of dopamine from fibres that run through the lateral hypothalamus enroute from the midbrain to the striatum and other forebrain structures. Although treatment with a large dose of reserpine removes 95% or more of the dopamine in the brain (based on analyses performed by Glen Baker), amphetamine facilitates the release of the remaining dopamine in amounts sufficient to restore normal function for a short time.

If electrocortical activation is necessary for the occurrence of directed Type 1 behavior it might no longer be possible to restore self-stimulation behavior in a reserpinized rat with an amphetamine injection if the rat was also treated with atropine or scopolamine. As we have seen, a combination of reserpine plus atropine or scopolamine abolishes all neocortical activation (and most hippocampal activation) and the activation patterns are not restored by a subsequent treatment with amphetamine (see Chapter 5).

I worked on this question with Mike Gutman, a 4th year undergraduate student who needed a project for his bachelor's thesis. Rats with chronically implanted lateral hypothalamic electrodes were trained to self-stimulate in an apparatus with two levers. Pressing one lever activated the brain stimulator and a counter (active lever); pressing the other lever activated only a counter (inactive lever). The rats rapidly learned to press the active lever at rates as high as 7,000 presses/hour while the inactive lever was largely ignored. If the functions of the active and inactive levers were interchanged, the rats began to press the formerly inactive lever within a few minutes.

Twelve to sixteen hours after a large dose of reserpine (10 mg/kg) the rats were cataleptic and could not be induced to self-stimulate. Following an additional treatment with amphetamine (1.0 mg/kg) they became spontaneously active within a few minutes and were soon self-stimulating at their normal rates. However, if atropine or scopolamine were administered together with (or prior to) the amphetamine, spontaneous motor activity was restored but self-stimulation did not occur. The rats just walked about aimlessly in the apparatus. Control experiments showed that atropine or scopolamine, alone or in combination with amphetamine, had only minor effects on self-stimulation in rats that had not been treated with reserpine. Therefore, these experiments demonstrate that total blockade of electrocortical activation is associated with a loss of hypothalamic self-stimulation behavior. It may be that electrocortical activation is essential for cerebral control of Type 1 behaviors such as walking to the active lever and pressing it with the forepaws[8].

Even as the paper describing these experiments on self-stimulation was in press, other ongoing work revealed that a combination of *p*-chlorophenylalanine and either atropine or scopolamine would completely abolish neocortical and hippocampal activation (see Chapters 5 and 6). This provided a much more selective way of abolishing such activation than the combination of reserpine plus atropine or scopolamine. Almost immediately I began an investigation of the behavioral effects of combinations of *p*-chlorophenylalanine and scopolamine or atropine. In part of this work, I was assisted by a second very talented 4th year undergraduate student, Clayton Dickson, who subsequently went on to do graduate work with Brian Bland at the University of Calgary.

Rats treated with large doses of *p*-chlorophenylalanine and scopolamine or atropine walked about without any obvious impairment and were at least as active as control rats (injected with the drug vehicle only) when placed in a large open field. The drugged rats swam as well as control rats and could easily climb up a vertical piece of wire mesh. Despite this, they displayed gross abnormalities in the control of behavior. They would repeatedly and without hesitation walk over the edge of the 35x35 centimeter elevated movement sensor platform (raised about 60 centimeters above the laboratory bench) used in the experiments on recording cortical electrical activity. While recording brain activity, I had to hold them by the tail to prevent them from falling. Similar behavior has been observed in chronic surgically decerebrate rats[4]. Attempts were made to train the rats to swim to a large visible platform in the center of a water-filled aquarium, a task which normal or vehicle-injected control rats perform perfectly after one or two trials. The *p*-chlorophenylalanine and scopolamine treated rats were completely unable to do this. They swam repeatedly in circles around the perimeter of the aquarium and if they happened to encounter the platform they would either push themselves away from it or climb up, run over the platform, and tumble into the water again on the other side. This gross impairment in behavior was not alleviated by training the rats prior to the drug treatment. The behavior of avoiding an electric shock by jumping out of a box was also severely impaired or totally abolished, even though the rats were capable of jumping. Similar very severe deficits in behavior were observed in rats given scopolamine several weeks after receiving injections of the serotonergic neurotoxin 5,7-dihydroxytryptamine into the brain stem sites from which serotonergic inputs to the cerebrum originate. Therefore, it is very likely that the behavioral syndrome produced by a combination of *p*-chlorophenylalanine and scopolamine is really due to concurrent blockade of central serotonergic and cholinergic muscarinic neurotransmission[9]. It is interesting that the behavioral effects of a combination of *p*-chlorophenylalanine and scopolamine were much more severe than the effects of either drug alone. For example, if rats were pretrained on the swim-to-platform test, neither of the two drugs, given alone, has any effect on subsequent performance. If the two drugs were combined, however, subsequent performance declined almost to zero. Total blockade of cerebral cortical activation, it seems, has a far more severe effect on behavior than a partial blockade.

The behavioral deficits produced by a combination of *p*-chlorophenylalanine plus scopolamine or atropine are not restricted to learned behavior. The effect on self-grooming is a clear example of this. Grooming behavior in rats is a highly stereotyped taxon-specific behavior which is difficult to modify by training. Therefore, it can be regarded as largely unlearned or instinctive. A rat usually begins an episode of grooming by sitting up to lick its forepaws and rub them over the mouth and snout. Subsequently, the paws are rubbed over the area above the eyes and over the ears and the back of the head. After a period of face washing, a rat typically performs a sudden change in posture and begins to lick its hind legs, flanks, or abdomen. The postural change is a Type 1 behavior (accompanied by hippocampal rhythmical slow activity and suppression of the late component of the transcallosal evoked potential) while face-washing and licking the fur are Type 2 behaviors which are accompanied by electrocortical patterns similar to those occurring during waking immobility. Therefore, I thought that central blockade of central muscarinic and serotonergic neurotransmission would disrupt grooming behavior by reducing or abolishing cerebral control of the postural changes involved in grooming but without having much effect on the stereotyped components of grooming such as face-washing.

The problem was approached by carrying out frame-by-frame analyses of videotapes of rats grooming themselves for 30 minutes after a brief immersion in water following treatment with either saline (control condition) or following treatment with scopolamine (experimental condition)[10]. Each rat was tested under both conditions and the order of the conditions was counterbalanced (approximately half the rats were tested with saline first and the others received scopolamine first) Frame by frame analyses permitted a detailed study of the duration of the various component behaviors of a grooming sequence and of the sequential structure of the behavior.

Some of the results of such experiments are shown in Table 8-1. The duration of head wash sequences (including washing the snout and behind the ears) and the percentage of head wash sequences that included a successful transition from washing the snout to washing behind the ears were not significantly affected by scopolamine. However, the total number of head wash sequences, the duration of body grooming sequences, and the percentage of head washing sequences that were followed by body grooming were all strongly reduced. The occurrence of successful postural changes, such as the change from head washing to body grooming was reduced by about 88 percent. Many of the behavioral abnormalities produced by scopolamine alone were further increased in rats that had previously sustained a loss of central serotonergic projections as a result of intra-brain stem injections of the neurotoxin 5,7-dihydroxytryptamine.

What does all this mean and why would anyone be interested in the details of how rats groom themselves? In reply one can say that the results indicate that central anti-muscarinic blockade, alone or in combination with destruction of serotonergic neurons, disrupts the sequential structure of grooming behavior primarily by disturbing the cerebral control of the postural adjustments that intervene between the successive occurrences of the stereotyped components of grooming such as head-washing and licking the fur of the body. Successfully

Table 8-1. Effects of scopolamine (5.0 mg/kg, s.c.) on grooming behavior in normal rats *(N=9)*

	Saline	Scopolamine
1. Total number of head wash sequences	30.9 ± 3.9	13.3 ± 1.5[a]
2. Percent head wash sequences that include wiping over the ears	46.3 ± 6.2	34.4 ± 8.2
3. Duration of head-wash sequences (s)	3.9 ± 0.6	3.9 ± 0.4
4. Total number of body groom sequences	26.4 ± 4.5	3.1 ± 1.2[a]
5. Duration of body groom sequences (s)	8.6 ± 0.8	1.4 ± 0.5[a]
6. Percent head-wash sequences followed by body grooming	52.5 ± 6.6	16.8 ± 5.5[a]

$p<0.01$, Wilcoxon test. From Robertson, B.J., Boon, F., Cain, D.P., and Vanderwolf, C.H. (1999). Behavioral effects of anti-muscarinic, anti-serotonergic, and anti-NMDA treatments: hippocampal and neocortical slow wave electrophysiology predict the effects on grooming in the rat. *Brain Research, 838*: 234-240, with permission from Elsevier Science.

executed postural adjustments were rare in the scopolamine-treated rats and in consequence: (a) head washing was initiated less often than normal (it requires a postural change from an all-fours position to a sitting-up position); (b) body grooming was initiated less often than normal (it requires a postural change from a sitting-up crouch with the paws near the head to a more erect posture with the paws and mouth in contact with the body); and (c) there was a reduction in the duration of body grooming sequences. This occurred because postural adjustments normally move the mouth and forepaws to a new position on the body every few seconds. If successful performance of such postural adjustments is reduced, the duration of body grooming sequences will also be reduced. In contrast to these effects, scopolamine did not alter the duration of head-wash sequences or the transition from washing the snout to washing behind the ears. In sum, anti-muscarinic and antiserotonergic treatments disrupt grooming sequences by interfering with the normal cerebral control of Type 1 behavior, leaving the Type 2 behavior relatively intact. This effect, quite probably, is due to blockade of electrocortical activation.

Implicit in this conclusion is the assumption that surgical removal of the cerebral cortex would interfere with grooming in the same way that anti-muscarinic treatment does. This turned out to be largely true. Body grooming was depressed in surgically decorticated rats and the probability of a transition from face-washing to body grooming was reduced from a normal value of 0.6 to 0.03[5].

Very similar effects were observed on grooming following large cerebral cortical lesions in the Mongolian gerbil. In addition, rhythmical drumming of the hind feet, an interesting Type 2 behavior occurring in Mongolian gerbils, was not affected by large cortical lesions that abolished ventral scent marking (a Type 1 behavior that consists of dragging the abdominal scent glands over objects).[11]

There are at least two reasons why an understanding of the brain mechanisms involved in grooming behavior should be of interest to students of the nervous system and of mammalian biologists in general. First of all, grooming is important in its own right, making an important contribution to survival. Grooming plays a role in temperature regulation: dirty matted fur is a poor insulator. Grooming behavior also provides a means of removing external parasites, and of cleaning and caring for injuries. Mutual grooming has an important social function in many mammals. Second, beyond any interest in grooming in itself, the behavior can be viewed as a model for other more frequently studied activities. A rat in a water maze, for example, must make numerous decisions about which direction to swim in, when to stop swimming, and so forth. Similarly, a rat engaged in face-washing must make numerous decisions about the precise postural change to be performed next or whether to stop grooming and walk away. It is therefore not surprising that treatments with antimuscarinic drugs or lesions of the hippocampus disrupt both grooming and adaptive behaviors in a water maze[12].

The animal experiments on the role of ascending cholinergic and serotonergic pathways in the control of cortical activation and behavior may clarify the basis of some neurological and psychiatric conditions in humans. It has long been known that patients suffering from Alzheimer's disease display an abundance of large amplitude slow waves (delta and theta waves) in the electroencephalogram even though they are fully awake and conscious. By 1990 it was also clear that brain cholinergic and serotonergic functions were seriously compromised in Alzheimer's disease. Consequently, human Alzheimer's patients and rats with combined cholinergic - serotonergic blockade have: (a) similar neurochemical deficiencies (although the deficit is presumably more selective in the rats); (b) a similar predominance of large amplitude irregular slow waves in the electrocorticogram; (c) a capacity for relatively normal motor activity; and (d) a profound loss of organized adaptive behavior. It seems likely that in man, as in the rat, loss of cholinergic and serotonergic cortical activation produces a severe dementia.

At first sight, very severe dementia in humans does not appear to be directly comparable to the behavioral syndrome produced by combined central serotonergic and muscarinic blockade in rats. Patients in a persistent vegetative state,[13] attributed by Jennett and Plum to very extensive cerebral damage, display a behavioral sleep-waking cycle and a variety of reflexive and spontaneous behaviors. Such patients may open their eyes, may turn the head and eyes toward a sound, may smile, grunt, moan, or scream, can cough and swallow, and may move the trunk and limbs about in a meaningless way. They do not walk, speak, or fixate objects visually.

In contrast to such human patients, decorticate or high decerebrate laboratory animals are generally able to walk about quite well. These facts suggest that the assumption of an erect bipedal form of locomotion by ancestral hominids several million years ago was associated with an extensive reorganization of the neural control of locomotion. It seems probable that walking is much more dependent on the neocortex in man than in rats or cats. Similarly, the evolution of speech in humans was apparently associated with the appearance of a neocortical control of vocalization which is not present in other mammals.[14] If account is taken of these differences in the brain organization of humans and laboratory animals, it appears that the persistent vegetative state in humans is closely analogous to the behavioral syndromes produced in quadripeds by decortication, high decerebration or central blockade of serotonergic and muscarinic cholinergic function.

Notes on Chapter 8

1. Lindsley, D.B., Schreiner, L.H., Knowles, W.B., and Magoun, H.W. (1950). Behavioral and EEG changes following chronic brain stem lesions in the cat. *Electroencephalography and clinical Neurophysiology, 2*: 483-498.
2. Sherrington, C.S. (1906). *The integrative action of the nervous system.* New Haven: Yale University Press.
3. Gregoire, S.E., and Smith, D.E. (1975). Mouse-killing in the rat: Effects of sensory deficits on attack behaviour and stereotyped biting. *Animal Behaviour, 23*: 186-191.
   Zeigler, H.P. (1983). The trigeminal system and ingestive behavior. In E. Satinoff and P. Teitelbaum (eds.) *Handbook of behavioral neurobiology, volume 6, Motivation*, New York: Plenum Press, 265-327.
4. Woods, J.W. (1964). Behavior of chronic decerebrate rats. *Journal of Neurophysiology, 27*: 635-644.
5. Vanderwolf, C.H., Kolb, B., and Cooley, R.K. (1978). Behavior of the rat after removal of the neocortex and hippocampal formation. *Journal of Comparative and Physiological Psychology, 92*: 156-175.
6. Lashley, K.S. (1950). In search of the engram. In: *Symposia of the Society of Experimental Biology, No. 4, Physiological mechanisms in animal behavior*, 454-482.
   Myers, R.E., Sperry, R.W., and Miner McCurdy, N. (1962). Neural mechanisms in visual guidance of limb movement. *Archives of Neurology, 7*: 195-202.

7. Vanderwolf, C.H. (2001). The hippocampus as an olfacto-motor mechanism: were the classical anatomists right after all? *Behavioural Brain Research, 127*: 25-47.
8. Vanderwolf, C.H., Gutman, M., and Baker, G.B. (1984). Hypothalamic self-stimulation: the role of dopamine and possible relations to neocortical slow wave activity. *Behavioural Brain Research, 12*: 9-19.
9. Vanderwolf, C.H. (1987). Near-total loss of 'learning' and 'memory' as a result of combined cholinergic and serotonergic blockade in the rat. *Behavioural Brain Research, 23*: 43-57.

   Vanderwolf, C.H., Baker, G.B., and Dickson, C. (1990). Serotonergic control of cerebral activity and behavior: models of dementia. *Annals of the New York Academy of Sciences, 600*: 366-383. It is interesting that several psychotomimetic drugs (cyclazocine, normetazocine, phenylcyclidine, and phenylcyclohexylamine) suppress both atropine-resistant hippocampal rhythmical slow activity and atropine-resistant neocortical low voltage fast activity in a selective manner [Vanderwolf, C.H. (1987). Suppression of serotonin-dependent cerebral activation: a possible mechanism of action of some psychotomimetic drugs. *Brain Research, 414*: 109-118] and also produce (at least in the case of cyclazocine) a marked increase in the impairment produced by scopolamine in swim-to-platform performance in rats. [Buckton, G., Zibrowski, E.M. and Vanderwolf, C.H. (2001). Effects of cyclazocine and scopolamine on swim-to-platform performance in rats. *Brain Research, 922*: 229-233]. Thus, in terms of large-scale cerebral electrophysiology and behaviour, these psychotomimetic drugs have effects similar to the effects of *p*-chlorophenylalanine. The basis of this similarity deserves further investigation.
10. Robertson, B.J., Boon, F., Cain, D.P., and Vanderwolf, C.H. (1999). Behavioral effects of anti-muscarinic, anti-serotonergic, and anti-NMDA treatments: hippocampal and neocortical slow wave electrophysiology predict the effects on grooming in the rat. *Brain Research, 838*: 234-240.
11. Ellard, C.G., Stewart, D.J., Donaghy, S., and Vanderwolf, C.H. (1990). Behavioural effects of neocortical and cingulate lesions in the Mongolian gerbil. *Behavioural Brain Research, 36*: 41-51. This paper developed from an undergraduate teaching laboratory exercise. Colin Ellard (a graduate student working with my colleague Mel Goodale in the years 1982-86), and Dwight Stewart (a graduate student working with me) were teaching assistants in an undergraduate laboratory course on the brain and behavior which I taught for many years. One of the laboratory exercises that students were asked to complete consisted of observation of the effects of large neocortical lesions in the Mongolian gerbil.

Gerbils were chosen for this primarily because they remain lively and active in the bright, noisy environment of a student laboratory, a situation that elicits prolonged periods of crouching immobility in rats. Colin and Dwight became quite intrigued by this project, and with the assistance of Steve Donaghy, who needed a project for an undergraduate thesis, they produced an interesting paper showing, among other things, that several Type 2 behaviors were less affected by cortical lesions than Type 1 behaviors. Grooming behavior was affected by cortical lesions in much the same way that it is in rats.

12. Cannon, R.L., Paul, D.J., Baisden, R.H., and Woodruff, M.L. (1992). Alterations in self-grooming sequences in the rat as a consequence of hippocampal damage. *Psychobiology, 20*: 205-218.

    Morris, R.G.M., Garrud, P., Rawlings, N., and O'Keefe, J. (1982). Place navigation impaired in rats with hippocampal lesions. *Nature, 297*: 681-683.

    Sutherland, R.J., Kolb, B., and Whishaw, I.Q. (1982). Spatial mapping: definitive disruption by hippocampal or medial frontal cortical damage in the rat. *Neuroscience Letters, 31*: 271-276.

13. Jennett, B., and Plum, F. (1972). Persistent vegetative state after brain damage. *Lancet, 1*: 734-737.

    Plum, F. (1991). Coma and related global disturbances of the human conscious state. In: A. Peters and E.G. Jones (eds) *Cerebral cortex, vol. 9, Normal and altered states of function.* New York: Plenum Press, pp. 359-428.

14. Myers, R.E. (1976). Comparative neurology of vocalization and speech: proof of a dichotomy. *Annals of the New York Academy of Sciences, 280*: 745-757.

Chapter 9

Whatever Became of the Ascending Reticular Activating System?

According to conventional ideas on the subject, activation of the electrocorticogram is due to activity in the ascending reticular activating system, a diffuse system originating in the brain stem and projecting to the "non-specific" nuclei of the thalamus, especially the intralaminar nuclei. These nuclei, in turn, are supposed to send fibers to all regions of the cortex. Activity in this ascending system is said to produce neocortical activation and, therefore, to play a pivotal role in the sleep-waking cycle, arousal, consciousness, and attention.

In contrast to this, the experiments described in the preceding chapters indicate that neocortical activation is due to activity in ascending cholinergic projections from the basal forebrain to the neocortex and also to ascending serotonergic projections from the brainstem to the neocortex. The serotonergic inputs are active only in correlation with the occurrence of Type 1 behavior but, under some conditions, the cholinergic inputs are active during waking immobility as well as during Type 1 behavior. The thalamus, like the neocortex, may be a target of activating inputs (thereby blocking the rhythmical discharges that produce large amplitude rhythmical waveforms in the neocortex) but the thalamus is not an important causal link in the processes leading to neocortical activation. Furthermore, it is proposed that although the activating inputs to the cerebral cortex play a key role in the cerebral control of behavior, they are not directly involved in sleep, waking, or arousal. Selective blockade of ascending activation produces a dementia-like syndrome but not sleep, stupor, or coma as would be expected on the basis of the conventional theory of the reticular activating system.

These two accounts of neocortical activation differ in many respects. One (or perhaps both) of them must be incorrect. Let us examine a number of well-established facts in this field to determine where the truth lies.

A major assumption of the conventional theory is that the pattern displayed in the electrocorticogram is closely related to the sleep-waking cycle. Slow waves and spindles are said to be characteristic of sleep while a low voltage mixed frequency record is said to be characteristic of alert wakefulness. Is this true?

The discovery of rapid eye movement sleep (REM sleep or active sleep) demonstrated that a low voltage mixed frequency electrocorticogram could be present for long periods in sleeping humans or sleeping laboratory animals[1]. It is also true that a low voltage mixed frequency electrocorticogram is normally present during stages 1 and 2 of human sleep and the corresponding pattern of sleep in laboratory animals. Non-rapid eye movement sleep (quiet sleep) in other words, is sometimes accompanied by so-called sleep spindles and large amplitude slow waves and sometimes accompanied by a low voltage mixed frequency record[2]. Conversely, waking animals may display large amplitude slow waves in the waking state. This phenomenon, referred to as "post-reinforcement synchronization" because it was initially observed that a cat drinking milk usually displays a non-activated electrocorticogram, was studied extensively 25-30 years ago[3]. Taken together, these facts demonstrate that the electrocorticogram may be in either an activated or a non-activated condition in both the waking state and sleep. This is not a novel or original conclusion. N. Kleitman, a pioneer in the field of sleep research stated that "it is clear that the EEG by itself not only fails to gauge the depth of sleep but the very presence of behavioral sleep."[4] This view was supported by M. Jouvet, another innovator in sleep research, who wrote that "in no case does the state of the corticogram allow us to presume whether an animal is asleep or awake"[5].

Another indication that neocortical activity has no direct relation to the sleep-waking cycle is the fact, first demonstrated in experiments in dogs by N. Kleitman in 1932, that decorticate animals continue to display relatively normal sleep-waking cycles. Bryan Kolb, Richard Cooley and I studied sleep in rats in which the entire neocortex, cingulate cortex, and hippocampal formation had been surgically removed. These animals spent much of the day time sleeping, often in a curled up, nose-to-tail posture, just like intact rats. At night they became active, walking, running, and rearing much of the time[6]. Human patients in whom the cerebral cortex has been very extensively damaged may also display sleep-wakefulness cycling together with an "overwhelming dementia[7]".

It is also quite inaccurate to say that the electrocorticogram is related to consciousness as was discussed in Chapter 4. Anesthetized or otherwise comatose animals or humans may display an activated electrocorticogram; wakeful conscious animals or humans may display abundant large amplitude slow wave activity[3,7]. For example, this condition is present in Alzheimer's disease in humans. Severely demented patients display abundant large amplitude slow waves (delta and theta waves) in the electroencephalogram during waking behavior[8].

Although these basic facts have been known for a long time, some authors, moved perhaps by a strong desire to find *some* aspect of cerebral electrophysiology that can be linked directly to consciousness, continue to make proposals linking brain activity to subjective awareness. For example, several authors have suggested that the 30-80 Hz oscillations (γ or gamma waves) of the electrocorticogram play a key role in perceptual awareness.

I thought it would be worthwhile to do a few experiments on neocortical γ waves[9]. Figure 9-1 illustrates the results of one of them. A waking unanesthetized rat displays continuous neocortical γ activity as long as the electrocorticogram maintains a low voltage fast or mixed frequency pattern. If large amplitude slow waves appear, for example, during quiet sleep or deep surgical anesthesia, γ waves occur only in bursts with each burst riding on top of the surface-positive peak of the large slow waves. The surface-negative peaks are associated with a suppression of γ activity. This might suggest that a burst-suppression pattern of γ activity is characteristic of unconsciousness were it not for the fact that a tail pinch during deep surgical anesthesia can produce continuous γ activity with an amplitude substantially greater than the γ activity occurring during normal waking (Figure 9-1), even though the rat shows no sign of behavioral arousal.

On the other hand, if a rat is treated with a combination of *p*-chlorophenylalanine and scopolamine, large irregular slow waves, much like those occurring during natural sleep, are present continuously even though the rat is awake and hyperactive. The surface-positive components of these waves too, have bursts of γ waves riding on their peaks while the surface-negative peaks are associated with suppressed γ activity just as in sleeping or anesthetized rats (Figures 9-2 and 9-3). Therefore neocortical γ activity can be in either a continuous or a burst-suppression pattern, both during wakefulness and during surgical anesthesia.

It seems to me that the conclusion is inescapable: the electrocorticogram is not closely linked to the sleep-waking cycle or consciousness and the cerebral cortex is not essential for either sleep or the waking state. The conventional account of the relation of cortical activation to behavior or mental activity is quite wrong.

If the view that the electrocorticogram or electroencephalogram is closely related to consciousness and the sleep-waking cycle is really quite wrong, then

Figure 9-1. Spontaneous field potentials in the sensori-motor cortex before (ABC) and during (DEF) anesthesia induced by urethane (1500 mg/kg. i.p.).(A.D) Surface-to-depth bipolar activity: deep negative up: frequency band 0.3-90 Hz: vertical bar represents 1.0 mV. (B.E) Similar to A.D. but the frequency band is 30-90 Hz and the vertical bar represents 0.1 mV. (C.F.) 30-80 Hz activity rectified and integrated in 1.0-s intervals. At "p", the rat is handled and stroked. In DEF, the rat displays no flexion reflex to pinching but there is a corneal reflex and a feeble pinna reflex. At "tp". a severe tail pinch produced a slight increase in the rate of respiration but no movement. Cortical activation was accompanied by high amplitude gamma activity. Figures 9-1, 9-2, and 9-3 are taken from Vanderwolf. C.H. (2000). Are neocortical gamma waves related to consciousness? *Brain Research, 855*: 217-224, with permission from Elsevier Science.

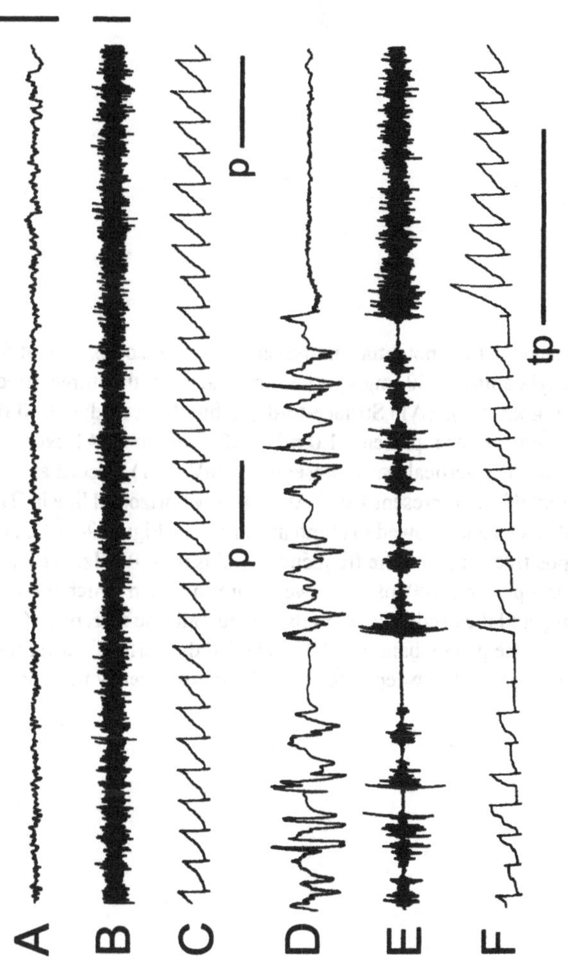
A
B
C
p
p
D
E
F
tp

Figure 9-2. Spontaneous field potentials in the sensori-motor cortex in a rat following treatment with *p*-chlorophenylalanine (500 mg/kg. i.p., on each of the three preceding days) plus scopolamine (5.0 mg/kg. s.c.). (A1) Surface-to-depth bipolar record; 0.3-90 Hz frequency band; deep negative up; vertical bar represents 1.0 mV. (A2) similar to A1 except that the frequency band is 30-90 Hz and the vertical bar represents 0.1 mV. (B1) Record as in A1 but at a lower speed. Thin horizontal lines represent 1.0 s. At the heavy horizontal line in B1 the rat was placed on a vertical metal screen and allowed to climb upwards (see Figure 9-3). (C. top) record as in A1 except that deep positive is up and the frequency band is 0.3-10 kHz: vertical bar represents 1.0 mV; horizontal bar represents 100 ms. Ten sweeps are shown in which the waves marked by the large white dot triggered the oscilloscope (pretriggered multiple sweeps). (C. bottom) Similar to C. top except that the frequency band is 30-100 Hz and the vertical bar represents 0.1 mV. Ten sweeps were taken after the 10 sweeps shown in C. top had already been recorded.

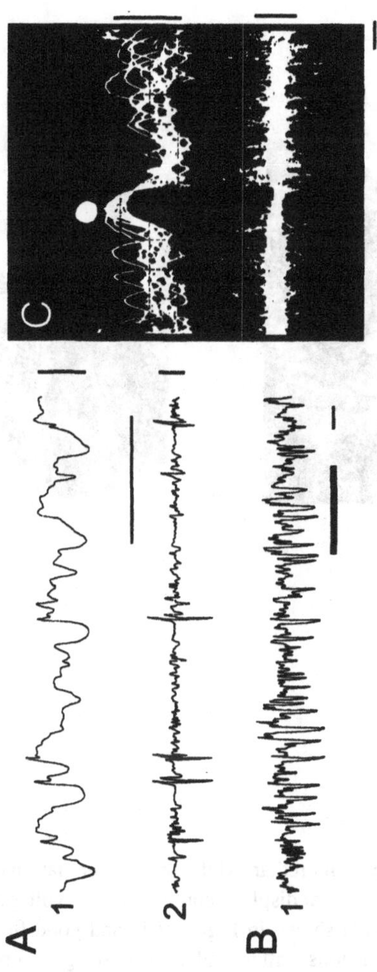
A
1
2
B
1
C

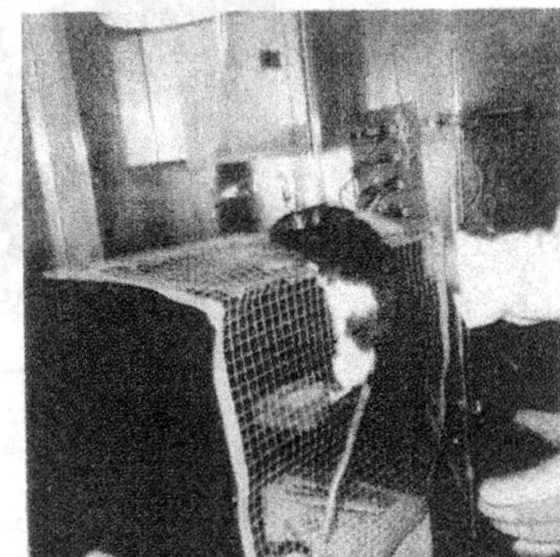

Figure 9-3. The behavioral effects of urethane (left) and a combination of *p*-chlorophenylalanine and scopolamine (right). (*Left*) This rat displays continuous low voltage fast activity (not shown but other records from this rat are shown in Figure 9-1) and good flexion, corneal, and pinna reflexes but does not right spontaneously after urethane (1000 mg/kg, i.p.) (*Right*) this rat (records shown in Figure 9-2) displays continuous high amplitude slow waves (corresponding to delta waves of the human electroencephalogram) together with vigorous spontaneous motor activity while climbing up a vertical wire screen.

why is it so widely accepted? I think the primary reason for this is that the early investigators of cerebral physiology in relation to behavior were satisfied with a very loose general correlation of cerebral activity and behavior as exemplified by the conventional classification of sleep into arbitrary stages with durations of many minutes. The data discussed in this book show that many aspects of cerebral field potential activity are related to behavior on a time scale of tens or hundreds of milliseconds. However, as long as people were content to relate cortical electrophysiology to such vague, imprecise concepts as consciousness and arousal, it was unlikely that anyone would notice its relations to the details of behavior. Even in the case of fairly obvious phenomena, people tend not to see things that they are not specifically looking for.

Is the theory that neocortical activation is mediated by a reticulo-thalamo-cortical pathway also wrong?

The honor of being the first to show that activation of the electrocorticogram is not mediated by thalamocortical projections belongs to J. Schlag and F. Chaillet of the University of California at Los Angeles. These investigators showed in 1963 that the neocortical activation produced by electrical stimulation of the intralaminar nuclei can be abolished by transverse cuts *caudal* to the thalamus which do not interfere with thalamocortical effects which pass *rostral* and *lateral* to the thalamus[10]. This means that the intralaminar region produces activation by a descending effect independent of thalamocortical projections. Years later, Dwight Stewart and I confirmed Schlag's and Chaillet's general result by showing that a combination of reserpine and scopolamine which abolished all traces of neocortical activation had no effect at all on various measures of thalamocortical transmission from the intralaminar nuclei or from specific sensory nuclei in the thalamus[11]. Further, there is an extensive series of publications showing that chronic lesions of the thalamus in both laboratory animals and humans do not abolish neocortical activation. For example, a 21-year old woman who survived prolonged cardiopulmonary arrest continued to display good electroencephalographic activation despite a severely atrophic thalamus (as determined at autopsy)[12,13].

There are many other findings that show that thalamocortical projections cannot be the cause of neocortical activation. For example, there is a wealth of evidence to show that release of acetylcholine or serotonin locally in the neocortex elicits activation of the electrocorticogram. The afferent systems that normally produce such effects cannot arise from the thalamus because the thalamus does not contain cholinergic or serotonergic neurons.

A major difficulty with the cholinergic-serotonergic theory of cortical activation as originally proposed is that activity in several other neurotransmitter systems also produce neocortical and hippocampal activation. For example, as I had observed myself (see Chapter 5), treatment with L-dihydroxyphenylalanine (the direct precursor of dopamine and norepinephrine) rapidly activated the electrocorticogram in a reserpine-treated rat. Since such activation was totally abolished by a subsequent systemic injection of atropine or scopolamine, I assumed that the activating effect was indirect, produced by dopamine (or norepinephrine) stimulating the cholinergic basal forebrain cells that project to the neocortex. Other evidence in support of this idea was the fact, first discovered by G. Pepeu of the University of Florence, that L-dihydroxyphenylalanine increased the release of acetylcholine from the neocortex.

Brain systems that depend on histamine as a neurotransmitter may also affect neocortical and hippocampal activity indirectly, acting via ascending cholinergic and serotonergic systems. Evidence favouring this possibility was obtained in experiments conducted by Philip Servos, a graduate student working with my colleague Mel Goodale in the years 1989-1994. Since Phil wanted to do a side-project of some sort (separate from research for his Ph.D.), I suggested a study of the electrocortical effects of α-fluoromethylhistidine, a drug which had been reported to reduce neuronal histamine to negligible values by inhibiting the enzyme histidine decarboxylase. The necessary assays of brain histamine levels were obtained by collaboration with L.B. Hough and his student K.E. Barke in the Department of Pharmacology and Toxicology at Albany Medical College in Albany, New York. Phil and I never actually met Hough and Barke: the entire arrangement was made by telephone, fax, and air shipment of frozen rat brains.

α-Fluoromethylhistidine, in common with H_1 receptor antagonists that penetrate the blood brain barrier, tends to promote sleep. Nonetheless Phil found that the drug had no obvious effect on either atropine-sensitive or atropine-resistant neocortical low voltage activity or on atropine-resistant hippocampal rhythmical slow activity. Therefore, it is likely that any effects that central histaminergic blockade may have on electrocortical activity are indirect and mediated by the ascending cholinergic and serotonergic projections.[14]

Whether or not all non-cholinergic and non-serotonergic means of activating the electrocorticogram are similarly indirect seemed like a splendid topic for a Ph.D. thesis. A promising candidate for this type of work appeared in the person of Hans Christian Dringenberg who arrived from the University of Lethbridge in the fall of 1991, strongly recommended by both Bryan Kolb and Ian Whishaw. Kolb's recommendation was especially succinct and to the point. "Take this one", he wrote. Hans enrolled in our recently established neuroscience program, receiving an MSc in 1993 and a Ph.D. in 1996.

Hans' experiments provided evidence that stimulation of a number of different brain systems produced neocortical activation indirectly by exciting either the basal forebrain cholinergic system or the brain stem serotonergic system[15]. He began with a series of experiments on the amygdala, a brain region in which electrical stimulation produces a clear-cut activation of the neocortex. This activating effect, readily demonstrable in urethane-anesthetized rats, was completely abolished by systemic administration of scopolamine, suggesting that it was mediated by known projections from the amygdala to the cholinergic cells of the basal forebrain. Evidence that this was so was obtained by recording from single neurons in the basal forebrain region. These neurons were classified into two main types. A^+ neurons fired spontaneously at a higher rate when the cortex was activated than when it was not activated. Such neurons are likely to be cholinergic and to project to the neocortex (see Chapter 7). A^- neurons fired at a higher rate when the neocortex was not activated than when it was activated. In a third group of neurons the firing rate was unrelated to the presence of neocortical activation. Most A^+ neurons were excited (i.e. increased their firing rate) by a single pulse of electrical stimulation of the amygdala but the firing rate of most A^- neurons was decreased by such stimulation (Figure 9-4). If, as seems likely, the A^+ neurons are mostly cortically-projecting cholinergic neurons and the A^- neurons are mostly inhibitory interneurons, then amygdaloid stimulation acts by exciting the cholinergic neurons and inhibiting the inhibitory interneurons. The cholinergic neurons then release acetylcholine in the neocortex producing a low voltage mixed frequency response (activation) in the neocortex. Additional evidence that this was actually happening was provided in experiments in which lidocaine (a local anesthetic) was injected unilaterally in the basal forebrain. Prior to a lidocaine injection, unilateral stimulation of the amygdala produced activation of both the right and left neocortex but after the infusion of lidocaine into the basal forebrain on one side, neocortical activation on that side was totally abolished (Figure 9-5). These facts indicate that the amygdaloid nuclei project bilaterally to the basal forebrain and that the neocortical activation produced by stimulation of the amygdala is dependent on the cholinergic projections to the neocortex from the basal forebrain.

Electrical stimulation of the locus coeruleus region also produced neocortical activation in Hans' experiments. Since the activating effect was abolished by systemic injections of scopolamine and since single unit activity in the basal forebrain was affected in much the same way by locus coeruleus stimulation as by amygdaloid stimulation, Hans and I concluded that the activation produced by stimulation of the locus coeruleus is mediated by an ascending excitatory effect on the basal forebrain cholinergic neurons. The direct noradrenergic projections from the locus coeruleus to the neocortex do not seem to be directly involved in neocortical activation.

Unlike the cases of the amygdaloid nuclei and the locus coeruleus, stimulation of the dorsal raphe nucleus (a source of ascending serotonergic fibres) or the superior colliculus seems to produce neocortical activation by an effect on ascending serotonergic projections. This was revealed by experiments

Figure 9-4. The effects of single pulse stimulation of the amygdala on the discharge of two types of cells recorded extracellularly in the basal forebrain of urethane-anesthetized rats. A: Example of the effect of stimulation (Stim.; 500 μA, 1 ms pulse) of the amygdala on the activity of a cell that discharges at higher rates during the presence of activation (*Activ.*; mean rate during activation is 17.5 spikes/s) than during synchronized (*Synchron.*; mean rate during synchronized slow waves is 5.3 spikes/s) activity in the neocortex. Cells of this type often are excited by amygdala stimulation. (50 sweeps are shown; spike count refers to the output of a window discriminator. Calibration, 20 ms). B: Example of the effects of amygdala stimulation (300 μA, 1 ms pulses) on a cell that discharges at lower rates during the presence of activation (mean discharge=1.5Hz) relative to synchronized activity (mean discharge=5.0 Hz) in the neocortex. Typically, the discharge of these cells is reduced or suppressed by amygdala stimulation. (25 sweeps are shown. Calibration: 400 ms). Figures 9-4 and 9-5 are from Dringenburg, H.C., and Vanderwolf, C.H. (1998). Involvement of direct and indirect pathways in electrocorticographic activation. *Neuroscience and Biobehavioral Reviews, 22*: 243-257, with permission from Elsevier Science.

A: Basal Forebrain Cell
(Activ. > Synchron.)

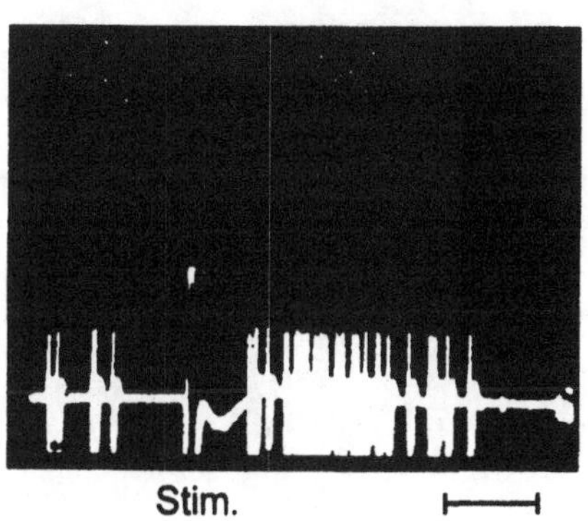

B: Basal Forebrain Cell
(Activ. < Synchron.)

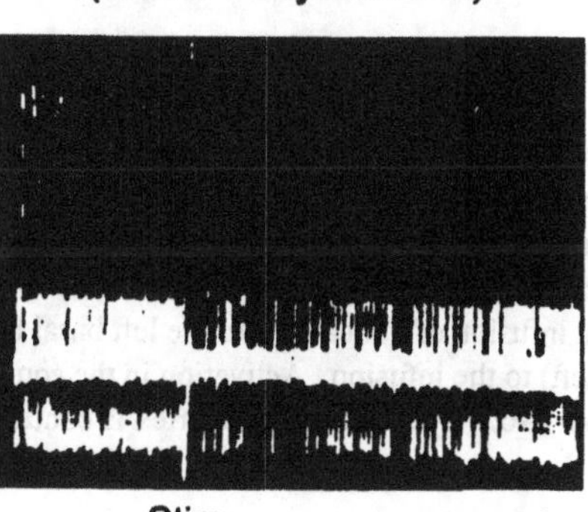

Figure 9-5. The effect of a unilateral lidocaine (1.0%) infusion into the basal forebrain of a urethane-anesthetized rat on neocortical activation during stimulation of the basal amygdala. Prior to the infusion, unilateral amygdala stimulation elicits activation in both cortical hemispheres. A saline infusion into the basal forebrain has no effects on activation during amygdala stimulation. However, a subsequent infusion of lidocaine into the left basal forebrain blocks activation in the neocortex ipsilateral (left) to the infusion. Activation in the contralateral (right) neocortex is not blocked by the infusion. Activity of the left neocortex is contaminated by stimulation artifacts. Calibrations: 0.5 mV; 1s.

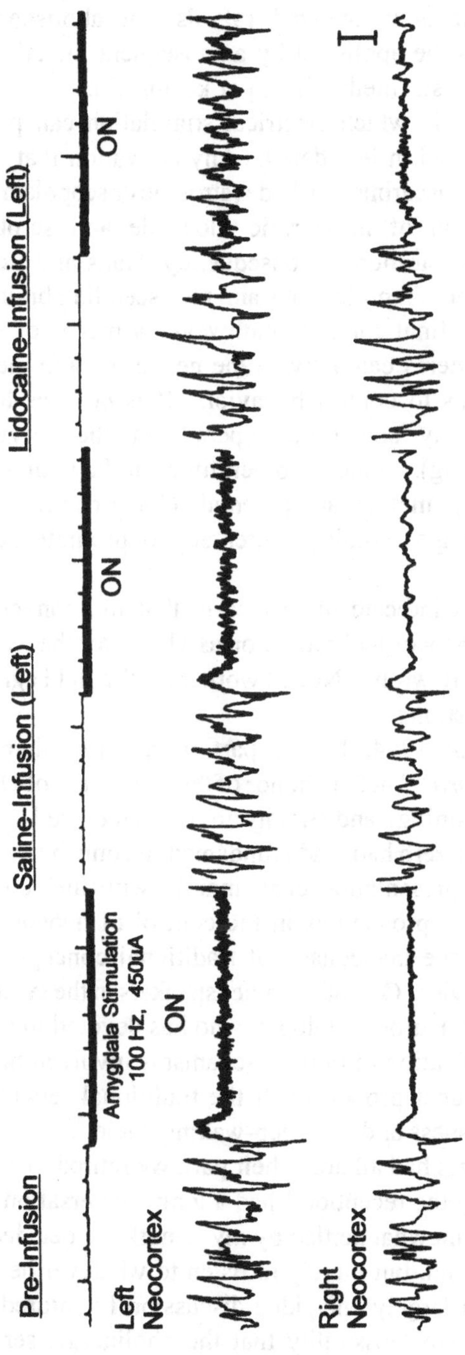
Pre-Infusion
Saline-Infusion (Left)
Lidocaine-Infusion (Left)
Amygdala Stimulation
100 Hz, 450uA
ON
ON
ON
Left
Neocortex
Right
Neocortex

showing that the neocortical activation produced by stimulation of these structures in urethane-anesthetized rats is not abolished by atropine or scopolamine but can be abolished by a subsequent injection of either of two serotonergic antagonists, methiothepin or ketanserin.

Other brain sites in which electrical stimulation can produce neocortical activation were studied in less detail. Any activation that was produced was always blocked by muscarinic blockade (atropine or scopolamine administration) or by a combination of muscarinic blockade and serotonergic blockade (methiothepin administration). Consequently, Hans and I concluded that the ascending basal forebrain projections and the ascending brainstem serotonergic projections serve as final common pathways by means of which a variety of functional brain systems can activate the neocortex and thereby make use of cerebral mechanisms to control behavior. It is perhaps by this means that cerebral activity may sometimes operate in the service of nutritional requirements (feeding), while at other times it facilitates reproduction (by controlling courtship, mating, and parental behavior), serves thermoregulatory requirements (seeking or building shelter, etc.) or promotes survival by avoiding dangerous predators.

I think the facts indicate quite clearly that the conventional theories of reticular control of neocortical activation as a basis for the waking state, arousal, and consciousness are wrong. Not all workers in the field agree, however, as the following story illustrates.

In the spring of 1994, I took part in a symposium on "*Cholinergic mechanisms in the brain*" held in honor of the retirement of Dr. John Szerb of the Department of Physiology and Biophysics of Dalhousie University in Halifax, Nova Scotia. John Szerb had made fundamental contributions to the topic of the symposium. My presentation dealt mainly with the role of the ascending forebrain cholinergic projections in the control of neocortical activation and behavior and with the inadequacy of traditional concepts related to cortical activation and behavior. One of the other speakers in the symposium was Mircea Steriade, an elderly neurophysiologist who has devoted much of his illustrious career to the investigation of brain mechanisms involved in sleep and waking. Steriade was an ardent proponent of the traditional reticulo-thalamo-cortical theory of consciousness and the sleep-waking cycle[16].

After the speakers had all done their part, we retired to a reception followed by a dinner. During the reception I had a long conversation with Steriade. He had clearly been somewhat nettled by my remarks. I decided not to debate any of the issues with him, but merely to listen to whatever he might have to say. After a lengthy but largely one-sided discussion I ventured to ask whether he thought there was any possibility that the cholinergic-serotonergic-behavior control theory might replace the the reticulo-thalamo-cortical-consciousness theory. He looked at me steadily, drew himself up to his full height, and declared in his resonant Rumanian-accented voice, "Absolutely null!"

I made no reply to this: it is not unlikely that he was right. Mere facts seem to count for nothing against a superficially simple and widely accepted theory. Soon afterwards we all went in to dinner.

Notes on Chapter 9

1. Dement, W., & Kleitman, N. (1957). Cyclic variations in EEG during sleep and their relation to eye movements, body motility, and dreaming. *Electroencephalography and Clinical Neurophysiology, 9*: 673-690.
2. Bergmann, B.W., Winter, J.B., Rosenberg, R.S., & Rechtschaffen, A. (1987). NREM sleep with low voltage EEG in the rat. *Sleep, 10*: 1-11.
3. Vanderwolf, C.H., & Robinson, T.E. (1981). Reticulocortical activity and behavior: a critique of the arousal theory and a new synthesis. *The Behavioral and Brain Sciences, 4*: 459-514.
4. Kleitman, N. (1963). *Sleep and Wakefulness.* University of Chicago Press, Chicago, IL, (see p. 30).
5. Jouvet, M. (1967). Neurophysiology of the states of sleep. *Physiological Reviews, 47*: 117-177 (see p. 119).
6. Vanderwolf, C.H., Kolb, B., and Cooley, R.K. (1978). The behavior of the rat following removal of the neocortex and hippocampal formation. *Journal of Comparative and Physiological Psychology, 92*: 156-175.
7. Plum, F. (1991). Coma and related global disturbances of the human conscious state. In: A. Peters and E.G. Jones (eds). *Cerebral cortex, volume 9, Normal and altered states of function*, Plenum Press, New York, pp. 359-428.
8. Fenton, G.W. (1986). Electrophysiology of Alzheimer's disease. *British Medical Bulletin, 42*: 29-33.
9. Vanderwolf, C.H. (2000). Are neocortical gamma waves related to consciousness? *Brain Research, 855*: 217-224.
10. Schlag, J., and Chaillet, F. (1963). Thalamic mechanisms involved in cortical desynchronization and recruiting responses. *Electroencephalography and clinical Neurophysiology, 15*: 39-62.
11. Vanderwolf, C.H., and Stewart, D.J. (1988). Thalamic control of neocortical activation: a critical re-evaluation. *Brain Research Bulletin, 20*: 529-538.
12. Kinney, H.C., Korein, J., Panigrahy, A., Dikkes, P., and Goode, R. (1994). Neuropathological findings in the brain of Karen Ann Quinlan. *New England Journal of Medicine, 330*: 1469-1475.
13. Additional references to relevant animal experiments and cases of human brain damage are given by: Dringenberg, H.C., and Vanderwolf,

C.H. (1998). Involvement of direct and indirect pathways in electrocorticographic activation. *Neuroscience and Biobehavioral Reviews, 22*; 243-257.

14. Servos, P., Barke, K.E., Hough, L.B., and Vanderwolf, C.H. (1994). Histamine does not play an essential role in electrocortical activation during waking behavior. *Brain Research, 636*: 98-102.
15. Detailed accounts of Hans' main experiments can be found in: Dringenberg, H.C., and Vanderwolf, C.H. (1996). Cholinergic activation of the electrocorticogram: an amygdaloid activating system. *Experimental Brain Research, 108*: 285-296, and in Dringenberg, H.C., and Vanderwolf, C.H. (1997). Neocortical activation: modulation by multiple pathways acting on central cholinergic and serotonergic systems. *Experimental Brain Research, 116*: 160-174.
16. Steriade, M. (1996). Arousal: revisiting the reticular activating system. *Science, 272*: 225-226.

Chapter 10

Olfactory Reactions in the Dentate Gyrus and Pyriform Cortex

By the winter of 1990-91 I began to feel that I had worked long enough on hippocampal rhythmical slow activity and neocortical activation. It seemed to me to be fairly well established that these electrographic reactions were dependent on ascending cholinergic and serotonergic inputs and were related to the cerebral control of behavior but could not be understood in terms of conventional psychological concepts.

The time had arrived to return to the fast wave bursts that I had seen in the hilus of the dentate gyrus during sniffing about 25 years earlier (see Chapter 2). Richard Cooley prepared several rats with electrodes in the dentate hilus and also in the olfactory bulb or in the olfactory mucosa to record the slow depolarization of the olfactory receptors in response to odors. I thought I should use a strong odor to maximise the chances of seeing an effect. Since there was a histological set-up for sectioning and staining rat brain in the same room as my recording set-up, I simply dipped a Q-tip in the xylene used to prepare sections for coverslipping and presented it to a rat's snout while electrographic recordings were being made. The first presentation produced a large negative deflection (depolarization) in the olfactory mucosa (appearing as a positive response in the olfactory bulb) but had no obvious effect on the dentate gyrus. However, after an additional 2-3 presentations, a well-defined burst of rhythmical waves of about 20 Hz (β-waves) began to appear in the olfactory bulb and dentate gyrus (Figure 10-1)[1]. Once established, this response appeared to persist as long as I

Figure 10-1. Activity in the olfactory bulb and the dentate gyrus during the presentation of various sensory stimuli. OB, deep layers of olfactory bulb; H, stratum oriens of CA1; D, site in or just below the granule cell layer of the dorsal blade of the dentate gyrus; 10-50 Hz, integrated 10-50 Hz activity from the dentate record; M, motor activity recorded by the platform sensor; S, presentation of toluene by means of a Q-tip. Time is in seconds. Voltage calibration, 1.0 mV. Records OB, H, and D are all monopolar, negativity up. Note that toluene produced (a) predominantly positive potentials in the olfactory bulb followed by a rhythmical wave burst of about 20 Hz, (b) no clear effect at an RSA-generating site in the hippocampus, and (c) a fast wave burst of about 20 Hz in the dentate gyrus. The rat made no visible behavioral response to the toluene on this occasion. Spontaneous activity recorded in the olfactory bulb was rather depressed as a result of repeated tests with toluene and xylene. 1-5: integrated dentate responses to various stimuli. 1, toluene on another trial; 2, xylene; 3, cedarwood oil; 4, firing a starter's pistol which produced a violent startle response followed by running; 5, room lights flicked on and off. Black triangles indicate approximate time of stimulus application. From Vanderwolf, C.H. (1992). Hippocampal activity, olfaction and sniffing: an olfactory input to the dentate gyrus. *Brain Research, 593*: 197-208 with permission from Elsevier Science.

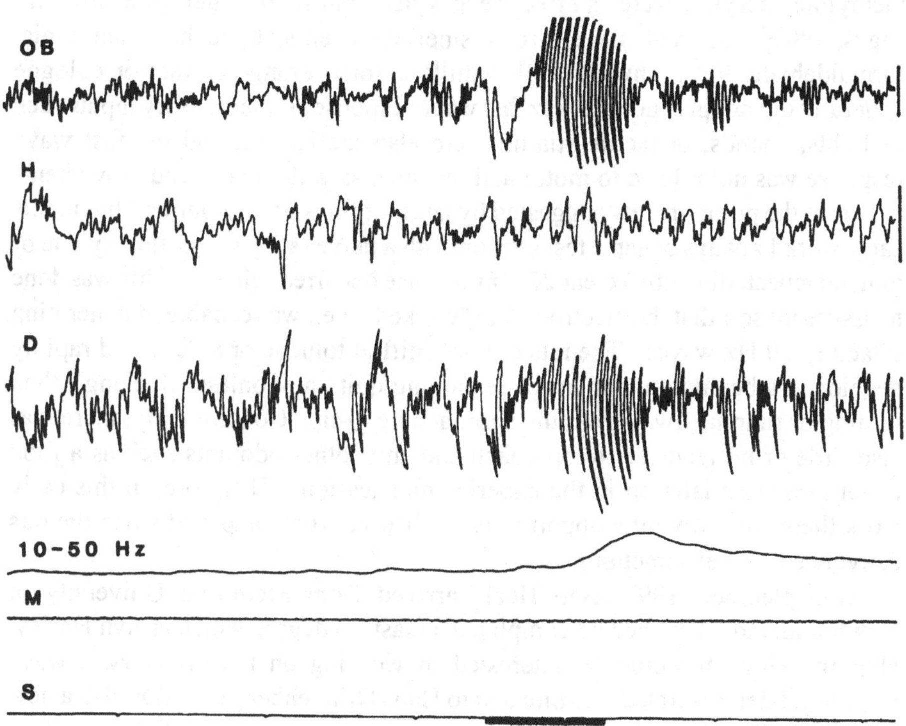
OB
H
D
10-50 Hz
M
S

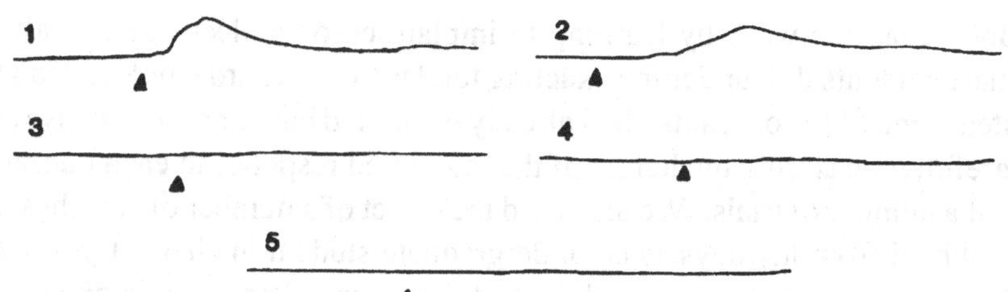
1
2
3
4
5

cared to continue presenting the Q-tip but it became more and more difficult to place the Q-tip near the snout because the rat became very proficient at dodging away. Clearly, the odor of xylene is aversive to rats. During the next few weeks I tried out a variety of different odorants. Toluene, diethyl ether, and methylmethacrylate were as effective as xylene but food, water, litter from rat cages, other rats, and some strong smelling chemicals such as ammonia, formaldehyde, 95% ethyl alcohol, vanilla extract, orange extract or cologne generally did not produce a 20 Hz fast wave response. Other sensory inputs such as lights, sounds, or tactile stimuli were also ineffective, and the fast wave response was not related to motor activity such as walking or head movement.

From the point of view suggested by later findings, it is important that in this early work I always began a test session with a naïve rat by presenting xylene or toluene repeatedly until a clear 20 Hz response occurred reliably. This was done to assure myself that the electrode site "worked", i.e., was capable of generating olfactory 20 Hz waves. The rats did not sniff at toluene or xylene and rapidly developed a behavioral pattern of remaining quite motionless (freezing), then turning or running away from the approaching Q-tip. Consequently, there was also little or no tendency to approach and sniff other odorants such as a food pellet presented later on in the experimental session. Therefore, in this early work there was very little opportunity to observe what happened when the rats actively sniffed at something.

In September, 1991, Bob Heale arrived from Memorial University in Newfoundland where he had completed a master's degree with Carolyn Harley. Hoping that Bob would be interested in working on the olfactory β-wave response, I demonstrated it to him and to Hans Dringenberg who was also a new student in the lab at that time. Without any overt prompting from me, Bob commented, "I'd like to work on that." I replied, "You've got it," and so his Ph.D. thesis topic in the Program in Neuroscience was settled in a matter of minutes.

Bob began his work by learning to implant chronic electrodes, as all new graduate students did, under the exacting tutelage of Richard Cooley. Bob then retested some of the odorants I had already examined but improved on my results by carefully measuring the height of the integrated response to each odorant on each of a number of trials. We also tried the effect of a number of stenches that I remembered from my days as an undergraduate student in chemistry – butyric acid, cadaverine, caproic acid, indole, and putrescine. Butyric acid gives rancid butter its sharp odor; caproic acid is a component of the characteristic odor of goats. Indole contributes to the odor of fresh feces while cadaverine and putrescine are diamines that contribute strongly to the odor of rotting flesh. Although these odorants were very effective in producing nausea in the people around the lab who smelled them [2], none of them produced a 20 Hz olfactory response in the olfactory bulb and dentate gyrus in the rats. Bob also tried the effect of squirting solutions of acetic acid, sucrose, sodium chloride, and quinine into the mouth of a number of rats[3]. This too failed to produce a 20 Hz response in the olfactory bulb and dentate gyrus, confirming the conclusion that this response was entirely olfactory and could not be elicited by stimulation of any other sense modality.

Around this time, my colleague Martin Kavaliers, who had a long–standing interest in olfaction and animal behavior, offered us a small quantity of 2-propylthietane, a substance which is the main component of the secretions of the anal scent glands of weasels. Bob found that this substance is as effective as the organic solvents toluene and xylene in eliciting a rhinencephalic 20 Hz response[3]. We also tried 2,4,5-trimethylthiazoline, a major component of the anal scent gland secretions of the red fox. This also produced a 20 Hz olfactory response in the olfactory bulb and dentate gyrus. Since weasels (or stoats) and foxes are predators of rats, these findings suggested to us that the 20 Hz rhinencephalic response might be related to the activation of a predator detection system. On this view, organic solvents such as diethyl ether, toluene and xylene are effective elicitors of the 20 Hz response because they mimic the action of the naturally occurring predator odors.

In September, 1994, Elaine Zibrowski arrived from Brandon University in Manitoba to begin graduate work with me in the Program in Neuroscience. Elaine soon discovered that the pyriform cortex at the base of the rat brain, like the olfactory bulb and dentate gyrus, developed large amplitude bursts of approximately 20 Hz waves in response to the odors of xylene, toluene, methylmethacrylate, 2-propylthietane, and 2,4,5-trimethylthiazoline. These waves were not elicited by visual, auditory, somatosensory, or gustatory stimuli or by the odors of ammonia, butyric acid, cadaverine or caproic acid. Their occurrence was not related to gross behavior such as locomotion or immobility and they were impossible to elicit if the nostrils were glued shut with a cyanoacrylate glue. (This was a terminal procedure performed shortly before the rats were anesthetized, killed, and the brain extracted for histological study). Therefore, as in the dentate gyrus and olfactory bulb, the pyriform 20 Hz or β-wave response is rather exclusively linked to a specific type of olfactory input[4].

The pyriform cortex displays an activation response which is rather similar to the activation response displayed by the neocortex. During quiet sleep, Elaine observed large amplitude irregular waves in the 1-20 Hz range (no attempt was made to determine if these waves are *always* present during quiet sleep) but if the rat was awakened, a lower amplitude higher frequency record appeared. Scopolamine hydrobromide produced a state in which the large amplitude irregular pattern of activity was continuously present in the pyriform cortex even when the rat was walking about vigorously. Therefore, the activation pattern of the pyriform cortex is quite possibly completely dependent on a muscarinic cholinergic input. Unlike the neocortex and the hippocampus, the pyriform cortex does not appear to receive a non-cholinergic input which produces activation in correlation with the occurrence of Type 1 behavior. This suggests that the function of the pyriform cortex with respect to behavior is quite different from the function of the neocortex and hippocampus. This topic requires much more extensive investigation.

The rat predator odors that we were using at this time, 2-propylthietane and 2,4,5-trimethylthiazoline, were synthetic products manufactured by Phero Tech Inc., of Delta, British Columbia. Predator odors had been shown to be of value in protecting orchards or recently reforested areas from small herbivores such as mice, voles, rabbits and hares. Consequently, there was a market for synthetic predator odors or other compounds that might repel small herbivores. David Wakarchuk, the Product Development Manager of Phero Tech, was quite interested in the 20 Hz olfactory response since it showed promise as a quick and relatively easy method of identifying new compounds that might deter small herbivores from feeding on seedlings or other valuable crops. We began a collaborative project, funded in part by the National Research Council of Canada. Phero Tech provided us with a variety of compounds previously shown (or suspected) to have the ability to suppress feeding in voles, snowshore hares, or other herbivores. At the same time Elaine and I, with the help of Tim Hoh, a graduate student working with my colleague Dr. Peter Cain, tested the effectiveness of a wider range of organic solvents to see if all members of this diverse group of compounds would elicit olfactory ß-waves[5]. We found that the organic solvents carbon tetrachloride, chloroform, 1,2-dimethoxyethane, *n*-heptane, mesitylene and methylcyclohexane as well as diethyl ether, methylmethacrylate, toluene, and xylene which had been tested earlier, were all very effective in eliciting an olfactory ß-wave response in both the pyriform cortex and the dentate gyrus. Commercially available gasoline, kerosene and turpentine were also effective. Curiously, two commonly used organic solvents, dimethylsulfoxide and N,N-dimethylformamide, failed to elicit a ß-wave response in our rats. A number of compounds that occur naturally in various plants were moderately to strongly effective in eliciting olfactory ß-waves in the pyriform cortex or dentate gyrus. These included: benzyl alcohol (found in jasmine and hyacinth; used in perfumes), camphor (found in a tropical tree; used to repel moths and externally as an antiseptic and counterirritant), carvacrol (found in spices such as thyme or summer savory), eucalyptol (found in eucalyptus trees; used to flavor some cough drops), and salicylaldehyde (found in some flowers; used in perfumery). It is apparent that a great variety of odorous chemicals can elicit an olfactory ß-wave response in rats. A similar situation is found in other chemical senses. Thus, compounds as varied as sucrose, glycerin, a dipeptide ester of aspartic acid and phenylalanine (aspartame), and lead acetate all taste sweet.

These results showed that a β-wave response can be elicited by effective odorants in the olfactory bulb, pyriform cortex and dentate gyrus. Ron Racine and his colleagues at McMaster University showed that the entorhinal cortex was also activated and that measurement of the delays in the appearance of the waves in the various structures is compatible with the activation of a multi-synaptic circuit running from the olfactory bulb to the pyrform cortex, then in succession to the entorhinal cortex and the dentate gyrus. Such a progression of activity was confirmed by Bob Heale and me in an experiment in which the projections from the entorhinal cortex to the dentate gyrus were surgically interrupted[6]. This procedure abolished the dentate gyrus β-wave response to the odor of toluene but did not affect the β-wave response in the olfactory bulb. Control lesions in the septal nuclei or in the amygdala did not have this effect. Whether a 20 Hz pattern of activity in this functional pathway should be understood as a sensory component of a predator avoidance mechanism is a question that requires further investigation. We found that isopentenylmethyl sulfide, a compound which, like trimethylthiazoline, is a component of the scent of foxes, will elicit an olfactory 20 Hz response in the pyriform cortex of both the rat and the meadow vole *Microtus pennsylvanicus* but that trimethylthiazoline is effective in the rat but not in the vole.[7] Thus, there appear to be species differences in the odorants that elicit the 20 Hz rhinencephalic response. Furthermore, we found that some compounds that deter rodents, rabbits, etc. from feeding do not produce the 20 Hz response. Since there are undoubtedly many different reasons why animals would avoid a particular odor, these facts do not disprove the hypothesis that the rhinencephalic 20 Hz response is related to predator detection but they do show that the situation is quite complicated.

The experiments on recording from the pyriform cortex had shown us that, in addition to the bursts of β-waves that could be elicited by specific odorants, there were prominent bursts of mostly 40-60 Hz activity (γ-waves) which occurred spontaneously in the sense that they occurred without any obvious eliciting stimulus. These waves had been studied extensively in freely moving animals by Walter Freeman and his associates at the University of California at Berkeley. Freeman found that the occurrence of the waves could be modified by various behavioral training procedures and concluded that they were related to one or more of arousal, motivation, attention, or expectancy. I decided to carry out an investigation of my own in the hope that something a bit more concrete and specific might emerge[8].

I began by comparing the occurrence of γ wave activity in the pyriform cortex during waking immobility and during spontaneous or induced locomotion. (Inducing locomotion means that I pushed the rat to make it walk forward). It

Figure 10-2. Pyriform gamma wave activity in relation to various patterns of respiration (rat #3934). R.PYR. 1-90, surface-to-depth bipolar record from right pyriform cortex, 30-90 Hz frequency band: MUC. 0.3-90, monopolar record (negative up, indicating inspiration) from the left olfactory mucosa, 0.3-90 Hz frequency band: Movement, output from a magnet-and-coil type movement sensor: R, continuous rhythmical breathing; A, apneic period; VSB, very slow breathing; INTEG. 30-80, right pyriform cortex activity, band-pass 30-80 Hz, rectified and integrated over 1.0 s intervals. Calibration, 1.0 mV. Note: (i) high levels of gamma activity during rhythmical breathing; (ii) lower levels of gamma activity during a 5-s period of apnea; and (iii) bursts of gamma activity associated with each breath during very slow breathing. Apart from the movements of breathing, the rat is completely immobile in a head-up, eyes-open posture. Figures 10-2 and 10-3 are from Vanderwolf, C.H. (2000). What is the significance of gamma wave activity in the pyriform cortex? *Brain Research, 877*: 125-133, with permission from Elsevier Science.

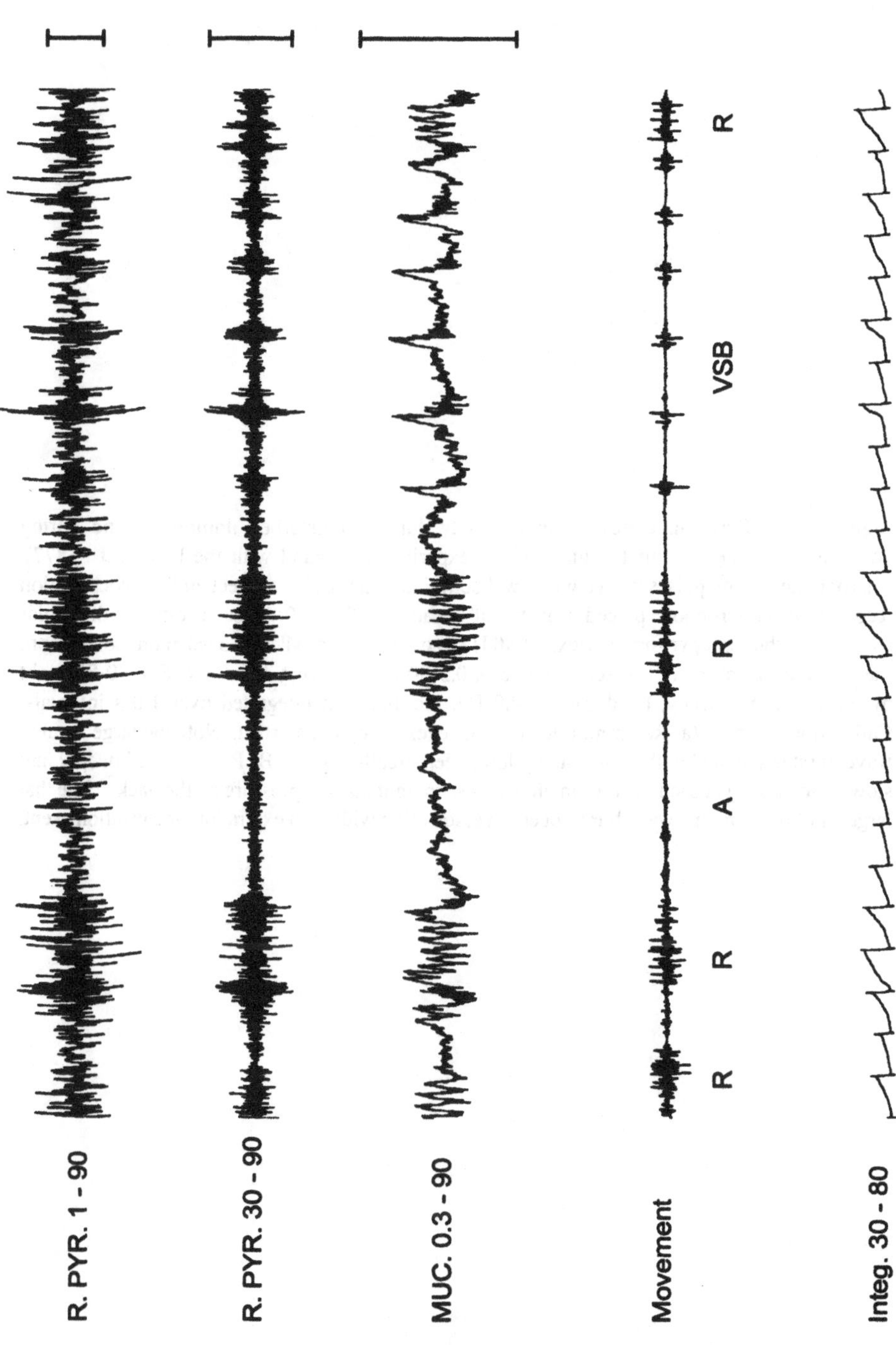
R. PYR. 1 - 90
R. PYR. 30 - 90
MUC. 0.3 - 90
Movement
R
R
A
R
VSB
R
Integ. 30 - 80

Figure 10-3. Pyriform cortex gamma activity during undisturbed immobility and during vocalization elicited by light contact of the experimenter's hand with the back (rat #3877). Microphone, output pulses from a window discriminator adjusted to detect audible vocalization recorded by a microphone placed near the rat's head: R. PYR. 30-90, surface-to-depth bipolar record from the right pyriform cortex, 30-90 Hz frequency band; MUC. 0.3-90, monopolar record (negative up) from the left olfactory mucosa, 0.3-90 Hz frequency band; INTEG. 30-80, right pyriform cortex activity, band pass 30-80 Hz, rectified and integrated over 1.0 s intervals. Calibration, 1.0 mV. (a) Rat immobile, head up, eyes open, undisturbed. Note the large gamma wave burst associated with an unusually long deep breath (sigh). (B) Rat breathed deeply and slowly, vocalizing at each expiration, in response to light manual pressure on the back. Note that large amplitude gamma wave bursts occur in association with each expiration-vocalization event.

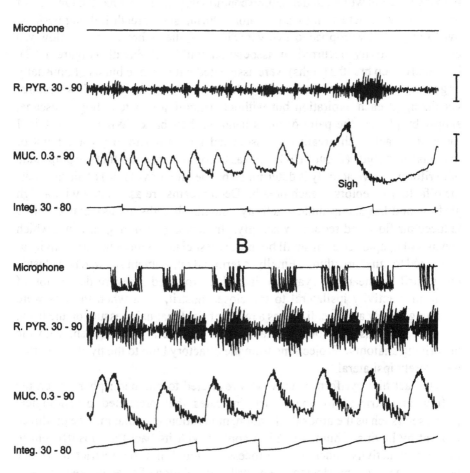
A
Microphone
R. PYR. 30 - 90
MUC. 0.3 - 90
Sigh
Integ. 30 - 80
B
Microphone
R. PYR. 30 - 90
MUC. 0.3 - 90
Integ. 30 - 80

was immediately apparent that pyriform γ activity is quite unlike hippocampal rhythmical slow activity: it is not increased by locomotion. It also became apparent that pyriform γ activity was very closely related to the pattern of breathing and its olfactory consequences which I monitored by recording the breath-by-breath depolarizations of the olfactory receptors in the olfactory mucosa. Rats often hold their breath for several seconds at a time when they are in an unfamiliar situation. This appears to be part of a generalized immobility or freezing reaction which, under natural conditions, presumably has the function of helping to avoid detection by a predator. During such breath-holding periods, pyriform γ activity dropped to a very low level. If the rat breathed very slowly, a burst of γ activity occurred in association with each breath (Figure 10-2). Unusually deep breaths (sighs) were associated with γ wave bursts of unusually large amplitude. Most rats can be induced to draw very deep slow breaths, vocalizing at each expiration but without struggling or attempting to escape, simply by placing the palm of one's hand on their back. When I tried this, I found that each such breath was associated with a pyriform γ wave burst of maximal amplitude (Figure 10-3). It was becoming apparent that the amplitude of pyriform γ wave activity is determined by the rate (or volume) of air flow past the olfactory receptors at each breath. Deep breaths are associated with a high air flow and large amplitude γ activity: smaller breaths are associated with a reduced air flow and reduced γ activity. In keeping with this, sniffing, which consists of a rapid series of small breaths, is associated with reduced γ activity as compared to quiet breathing. Finally, a terminal experiment showed that closing one nostril by means of cyanoacrylate glue produced a severe depression of pyriform γ activity ipsilateral to the closed nostril, even when the rats were breathing deeply and vocalizing as a result of light manual pressure on the back. The fact that this reduction in γ activity is ipsilateral is consistent with the underlying anatomy. Projections from the olfactory bulb to the pyriform cortex are entirely ipsilateral.

The fact that pyriform γ activity is restricted to one hemisphere when the airflow is restricted to one nostril indicates that presumed psychological processes (such as the arousal, attention, motivation, etc., that may be produced in a rat by having a human hand in contact with its back) are insufficient to produce γ activity since such processes would not be restricted to one hemisphere merely by closing one nostril. Stimulation of olfactory receptors by high rates of airflow is essential. Further, the fact that sighing is associated with large amplitude bursts of γ activity suggests that arousal, attention, motivation, etc. are largely irrelevant to pyriform γ activity since sighing is not usually regarded as a sign of such psychological activities. All these observations are consistent with a conclusion made 60 years ago by E.D. Adrian of Cambridge University in England, who found, in experiments in anesthetized animals, that pyriform γ activity is elicited in proportion to the rate of airflow past the olfactory mucosa. Complex psychological hypotheses are not adequate to account for the occurrence of this wave form. Nonetheless, training animals or subjecting them to various non-olfactory stimuli may affect pyriform γ activity because these procedures alter the ongoing patterns of breathing and air flow.

It was natural to wonder whether the occurrence of γ activity in the dentate gyrus would be like γ activity in the pyriform cortex. I set about investigating this with a fresh series of rats with electrodes implanted in the olfactory mucosa and in the hilus of the dentate gyrus. It soon became apparent that the dentate gyrus is quite different from the pyriform cortex. Large amplitude bursts of activity occurred in the hilus whenever a rat sniffed at something possessing an odor. Anything would do it seemed; my fingers, another rat, a rubber stopper, a pencil, the bristles of an old brush and so forth (Figure 10-4)[9]. Rats usually sniff at a novel object once for a few seconds but ignore the object on later presentations. In association with this, γ activity bursts were prominent in the dentate gyrus on first presentation but were much reduced or absent on later presentations. One exception to this was another rat. A rat will sniff at an unfamiliar rat very vigorously for seconds or minutes with high levels of dentate γ activity present as long as vigorous sniffing persists.

If one delivers a single brief electrical pulse of adequate strength to the perforant path, which contains projections from the entorhinal cortex to the dendrites of the dentate granule cells, an excitatory effect is obtained, as originally shown by Per Andersen and his associates in Oslo, Norway. The electrical events that are observed via an electrode placed in the dentate hilus are an initial positive potential, reflecting depolarization of granule cell dendrites, followed by a brief negative potential reflecting the mass discharge (population spike) of the bodies of the granule cells. To test the effect of sniffing on this series of events, I compared the effects of perforant path stimulation during quiet breathing and during the presence of vigorous sniffing. The result was clear. The population spike was consistently larger during sniffing than during quiet breathing, indicating that granule cells are more excitable when sniffing is occurring.

If dentate γ activity is correlated with sniffing behavior it could mean either: (a) that it is related to the motor activity (i.e. rapid small inhalations, vibrissae movement); or (b) that it is related to an afferent olfactory input. I suspected the latter because I had noticed that sniffing behavior when there was little or no olfactory input (sniffing in the air away from any odorous object; sniffing a thoroughly washed and air dried glass beaker) had little tendency to produce γ wave activity in the dentate gyrus.

A more definitive result was obtained by anesthetizing the rats with urethane. Under these conditions, blowing scented air into a nostril evoked a clear γ

Figure 10-4. Activity of the olfactory mucosa and dentate gyrus in a rat sniffing at a pencil (rat #3971). *Mucosa*, left olfactory mucosa, 0.3-90 Hz frequency band; *Movement*, output (1-90 Hz) from a magnet-and-coil arrangement attached to a light platform on which the rat was standing; *L. Dentate*, *R. Dentate*, 30-90 Hz frequency band of activity from the left and right dentate gyrus, respectively; *30-80 Hz*, 30-80 Hz activity from the left dentate gyrus rectified and integrated over 1.0 s intervals. Mucosa and dentate gyrus records are monopolar, negative up, with an indifferent placed in the skull over the cerebellum. Calibration: 1.0 mV, 5.0 s. At the heavy bar (marked S) the eraser end of a pencil was held near the rat's snout, provoking an increase in the frequency of the rhythmical respiratory potentials of the olfactory mucosa, accompanied by visible sniffing, head and vibrissae movement, plus a stepping movement (large deflection in the movement sensor record). The sniffing is associated with a bilateral high amplitude burst of gamma waves in the dentate gyrus. Note also that a brief apneic period (P) is not associated with any obvious change in dentate activity. From Vanderwolf, C.H. (2001). The hippocampus as an olfacto-motor mechanism: were the classical anatomists right after all? *Behavioral Brain Research, 127*: 25-47, with permission from Elsevier Science.

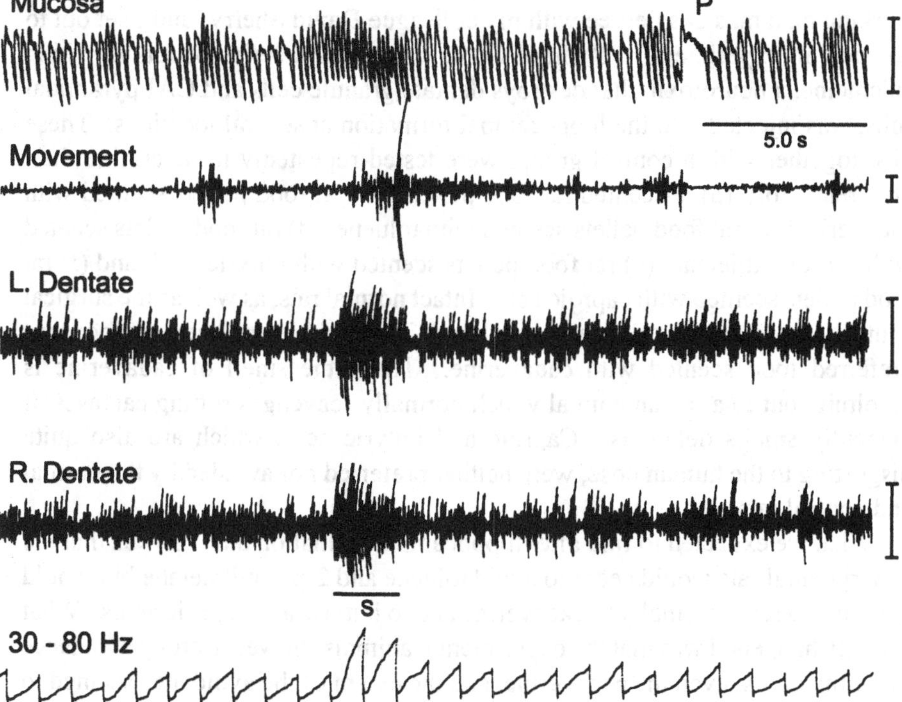
Mucosa
P
5.0 s
Movement
L. Dentate
R. Dentate
S
30 - 80 Hz

activity response in the dentate gyrus even though the rat was quite incapable of sniffing. Therefore, the γ activity response of the dentate gyrus when a rat sniffs at something appears to be a sign of excitation of dentate granule cells by an olfactory input.

The experiments on recording electrical activity in the dentate gyrus suggest strongly that this part of the brain is dominated by olfactory input, a conclusion already reached on anatomical grounds by the great Spanish neurobiologist Ramon y Cajal a century ago. More definite evidence of this would be obtained if it were possible to demonstrate changes in the olfactory control of behavior after injury to the dentate gyrus. Bob Heale, Karin Petersen (a graduate student working on a master's degree with my colleague David Sherry) and I set out to do such an experiment[10]. Richard Cooley prepared a series of rats in which colchicine, a neurotoxin that destroys dentate granule cells and CAI pyramidal cells, was injected into the hippocampal formation at several locations. These rats, together with a control group, were tested repeatedly to determine their preference for: (a) unscented rat food pellets; (b) rat food pellets scented with cadaverine; (c) rat food pellets scented with toluene; (d) rat food pellets scented with 2-propylthietane; (e) rat food pellets scented with butyric acid; and (f) rat food pellets scented with caproic acid. Intact normal rats, as well as the surgical control rats, avoided toluene-scented and 2-propylthietane-scented food but preferred food scented with cadaverine. To us the smell of cadaverine is revolting, but to a rat, an animal which normally scavenges rotting carcases, it evidently smells delicious. Caproic and butyric acid, which are also quite disgusting to the human nose, were neither preferred nor avoided by the normal and control rats.

What we expected to find after hippocampal formation damage was that the experimental rats would cease to avoid toluene and 2-propylthietane but would continue to react normally to cadaverine and to butyric and caproic acids. What actually happened was that the experimental animals showed a strong avoidance of *all* the odors, even the cadaverine, and the extent of the avoidance seemed to be correlated with the severity of hippocampal damage. We also tested the rats on the tendency (shown by normal rats) to avoid food flavored with quinine and prefer food flavored with sucrose. Since the hippocampal damage seemed to have no effect at all on such gustatory control of feeding behavior, we concluded that the changes in behavior that we observed were specific to the olfactory control of behavior.

Although the olfactory effects that we observed after hippocampal damage were not what we had expected, they are, nonetheless, consistent with the other evidence indicating that the hippocampal formation plays an important role in the olfactory control of behavior. Since the hippocampal formation also plays a role in the control of motor activity (see Chapters 2, 3 and 8) it can reasonably be regarded as an olfacto-motor region of the brain. The dentate gyrus is the main input zone (corresponding to layer 4 of the neocortex) while the pyramidal cells of Ammon's horn provide the output (corresponding to layer 5 of the neocortex). The nature of this olfacto-motor control awaits further experimental analysis.

It is well established that sensory inputs other than olfaction can influence hippocampal activity. For example, almost any type of sensory stimulus can elicit rhythmical slow waves in the hippocampus. It appears that olfactory inputs differ from other sensory inputs in having an oligosynaptic access to the dentate gyrus and in having a unique capability of eliciting a fast wave response in the dentate gyrus. The traditional view that the hippocampal formation is part of the rhinencephalon seems to be essentially correct.

From a historical point of view, it is interesting that Ramon y Cajal concluded that the hippocampal formation was primarily an olfactory structure on the basis of his discovery of a multisynaptic pathway from the olfactory bulb to the dentate gyrus and Ammon's horn. These findings were ignored for many years by those who hoped to show that psychological processes such as memory or emotion are localized in the hippocampus in some way. This appears to have been a mistake[9].

It is widely recognised that Ramon y Cajal had an amazing talent for getting things right the first time. During the 1963-64 academic year when I was working at the Institut für Hirnforschung (Brain Research Institute) in Zurich, Switzerland, the director, Konrad Akert, regaled the laboratory staff at lunch one day with stories about the Spanish neuroanatomist. He concluded with a word of advice: "If you think you have made an original discovery in neuroanatomy, the first thing you should do is read the relevant section of Ramon y Cajal's work to see if you got it right."

During the "olfactory phase" of my research, there were extensive changes in the activity of my lab. Bob Heale successfully completed a Ph.D. in neuroscience (1994), returned to Memorial University as a medical student, and eventually took up a career in neurosurgery. Elaine Zibrowski, having completed an MSc. in neuroscience (1996), took advantage of the current boom in opportunities in the field of epidemiology by obtaining an MSc. in that field, even as she continued to work part-time on our experiments on olfaction. When she graduated, she had no difficulty in obtaining a position in the mental health and drug addiction field. Richard Cooley, forced to retire by a medical condition after 28 years in the lab, was replaced by Francis Boon in 1998. Although I took no new graduate students after 1997, I continued to work in the lab by myself until I reached the mandatory retirement age of 65 in 2000. In the spring of 2001, I closed the lab and began writing this book.

Notes on Chapter 10

1. The increase (sensitization) in rhinencephalic responsivity with repeated exposure to an odor was described more fully in: Vanderwolf, C.H., and Zibrowski, E.M. (2001). Pyriform cortex β-waves: odor-specific sensitization following repeated olfactory stimulation. *Brain Research, 892*; 301-308.
2. The fact that there were only one or two informal complaints is an eloquent testimonial to the tolerance displayed by my colleagues and their students and technicians.
3. Heale, V.R., Vanderwolf, C.H., and Kavaliers, M. (1994). Components of weasel and fox odors elicit fast wave bursts in the dentate gyrus of rats. *Behavioral Brain Research, 63*: 159-165.
4. Zibrowski, E.M., and Vanderwolf, C.H. (1997). Oscillatory fast wave activity in the rat pyriform cortex: relations to olfaction and behavior. *Brain Research, 766*: 39-49.
5. Zibrowski, E.M., Hoh, T.E., and Vanderwolf, C.H. (1998). Fast wave activity in the rat rhinencephalon: elicitation by the odors of phytochemicals, organic solvents, and a rodent predator. *Brain Research, 800*: 207-215.
6. Heale, V.R., and Vanderwolf, C.H. (1999). Odor-induced fast waves in the dentate gyrus depend on a pathway through posterior cerebral cortex: effects of limbic lesions and trimethyltin. *Brain Research Bulletin, 50*: 291-299.
7. Vanderwolf, C.H., Zibrowski, E.M., and Wakarchuk, D. (2002). The ability of various chemicals to elicit olfactory β-waves in the pyriform cortex of meadow voles (*Microtus pennsylvanicus*) and laboratory rats (*Rattus norvegicus*). *Brain Research, 924*: 151-158.
8. Vanderwolf, C.H. (2000). What is the significance of gamma wave activity in the pyriform cortex? *Brain Research, 877*: 125-133.
9. Vanderwolf, C.H. (2001). The hippocampus as an olfacto-motor mechanism: were the classical anatomists right after all? *Behavioural Brain Research, 127*: 25-47.
10. Heale, V.R., Petersen, K., and Vanderwolf, C.H. (1996). Effect of colchicine-induced cell loss in the dentate gyrus and Ammon's horn on the olfactory control of feeding in rats. *Brain Research, 712*: 213-220.

Chapter 11

The Mind and Behavioral Neuroscience

Readers who have made their way through the preceding 10 chapters will have noted that none of the electrical events described in the neocortex, hippocampal formation and pyriform cortex have been related to mental activities. In fact, a good deal of evidence is presented to show that various aspects of brain electrical activity *are not* organized in accordance with conventional psychological ideas. A critical reader may well ask: "If the mind is not located in the cerebral cortex, then where is it?"

We must confront a fundamental question: what is behavioral neuroscience about? For many people, the ultimate problem of neuroscience is the neural basis of mental activity. According to this viewpoint, investigators should attempt to discover the neural basis of perception, attention, memory, emotion, motivation, cognition, etc. I will argue that this view is based on false assumptions and that it gives rise to bad science. Much of the conventional lore concerning the mind is a set of ancient hypotheses masquerading as facts. Further, I will argue that the problem of the mind is not soluble by contemporary science but that we may be able to make advances if we focus attention on the general problem of how the nervous system generates behavior. It is essential to be clear about this. The term "behavior" refers primarily to the postures and movements displayed by animals, including humans. Terms such as motivation or cognition refer to a universe of discourse quite distinct from behavior.

Having a long-standing interest in what one might refer to as the philosophical basis of neuroscience, I had hoped eventually to write a review summarizing the results of empirical studies relevant to the question of the mind and also to discuss the problems inherent in attempts to relate traditional mental concepts to brain activity. A marvellous opportunity to do this was presented by Brian and Cheryl Bland and Ian Whishaw who organized a symposium in Kananaskis, Alberta, in 1996 to commemorate my 60th birthday. The proceedings of the symposium were published as a festschrift in *Neuroscience and Biobehavioral Reviews* (volume 22, #2, 1998). The following discussion is loosely based on the ideas discussed in the review paper I prepared for this festschrift but also includes some additional topics related to the application of mentalistic concepts to neuroscientific research[1].

I. What is the Mind?

Isaac Newton (1642-1727), having discovered that a narrow beam of sunlight passing through a prism is dissociated into a series of bands of different colors, wrote: "The homogeneal Light or Rays which appear red, or rather make Objects appear so, I call Rubrifick or Red-Making; those which make Objects appear yellow, green, blue, and violet, I call Yellow-making, Green-making, Blue-making, Violet-making, and so of the rest... For the Rays to speak properly are not coloured. In them there is nothing else then a certain Power and Disposition to stir up a Sensation of this or that Colour"[2].

Thus, Newton, one of the giants of the scientific revolution that has transformed our world in the past four centuries, drew a clear distinction between two aspects of reality: (1) the objective physical world; and (2) our subjective experience of this world. If one accepts the conclusion, as many thinkers have, that our subjective experience of external reality is an entirely different thing from that reality itself, one is naturally led to some puzzling questions. What is the nature of subjective reality? Can we learn more about it, perhaps by adopting some of the methods of the natural sciences?

Rene Descartes (1596-1650)[3], argued that he could doubt the existence of the perceived world but could not doubt the existence of his own thoughts. He had, he said, dreamt that he was "dressed and seated near the fire, whilst in reality I was lying undressed in bed!" Further, there are no certain indications, he said, by which we may clearly distinguish wakefulness from sleep. Was it possible that he might really be dreaming when he believed himself to be awake? Consequently, external reality might be an illusion but it was impossible for him to doubt that he was having a subjective experience, regardless of whether it corresponded to reality or not. This line of reasoning led Descartes to the conclusion that the subjective world can be known directly and immediately while the external world can be known only by indirect means and by making inferences. "I see clearly that there is nothing which is easier for me to know than my mind", he wrote. If Descartes was right about this, we should be in possession of a great deal of detailed knowledge about the mind. What do we actually know about it?

A prominent feature of ideas about the mind is that it is a composite of different faculties or subprocesses or, alternatively, that it is a single entity capable of existing in many different states. People commonly discuss human experience and behavior in such terms as: anger, attention, belief, benevolence, cognition, contempt, desire, emotion, envy, esteem, grief, hatred, humility, humor, imagination, indignation, joy, love, lust, malice, memory, motivation, opinion, pain, perception, pity, pleasure, pride, respect, satisfaction, sensation and so forth. There seems to be no definite upper limit to the number of such terms.

The concept that the mind or psyche is a composite consisting of numerous subprocesses is very ancient. A detailed account of perception, emotion, memory, reason, imagination, thinking, etc., which is broadly similar to modern discussions of the same topics can be found in the writings of Aristotle (384-322 BC)[4]. Aristotle occupies a unique position in the history of natural science. His interests were broad, including what we now consider to be astronomy, physics, chemistry, zoology, botany, anatomy, physiology, animal behavior and psychology. Although he undoubtedly deserves great credit for his pioneering efforts to achieve a rational understanding of the natural world, many of his theories seem rather bizarre to us now. Aristotle believed that the most fundamental type of movement is circular and that linear movements are derived from circular motion. He thought that falling objects move at a constant velocity and that the material world is made up of four elements; fire, water, earth, and air. He did not have a coherent theory of light or vision, or of physiological activities such as breathing or digestion. Aristotle's psychology, however, is similar to its modern counterpart in many ways. Reason is said to be distinct from (and often opposed to) emotion or passion; thought always involves mental images, and thought proceeds by a process of association of ideas. Memory is compared to a physical information storage device. Aristotle favoured a comparison of memory to the effect of a signet ring pressed into wax: modern psychologists speak of encoding and retrieval, comparing human memory to computer memory. The details differ, but some essential features of the comparison are the same. All this is truly curious. We could conclude either that: (a) Aristotle was a failure in physics, chemistry, and physiology but a brilliantly successful psychologist; or (b) Aristotle's psychology is probably no better than his physics, chemistry, or physiology but that psychology has made no fundamental advance in analysing the mind in over two thousand years. Should we accept the conclusions about the psyche or mind drawn by Aristotle, Descartes and their many followers?

II. The Mind is not Open to Introspection

Descartes' ideas seem to suggest that the detailed nature of the mind should be obvious on simple introspection (the mind examining itself directly without the use of the external sense organs). However, a serious attempt to develop a science of mind based on the supposed method of introspection around 1880 lead to few results and considerable controversy because different observers could not agree. By 1910 introspective research on the mind had been largely abandoned[5].

There is much evidence that the brain processes responsible for our behavior are not in any sense open to introspective examination. If we ask people to multiply two numbers in their head or tell us the name of the capital of Sierra Leone they may provide the correct answer but they cannot tell us how the problem was solved. An electronic calculator or computer contains complex precisely designed circuitry which is well understood (by some people). Perhaps there are somewhat analogous circuits in our own brain but "introspection" tells us nothing about them. Again, we can tell at a glance whether a coffee cup is within reach or not but "introspection" does not tell us how we know this. Many years of research were required to discover the cues that provide information about distance such as retinal disparity, convergence, textural gradients, interposition, movement parallax, brightness, size, and linear perspective. As K.S. Lashley (1890-1958) once put it "No activity of mind is ever conscious"[6]. Therefore, the conventional taxonomy of mental processes cannot be based on introspection.

III. The Conventional Subdivisions of the Mind are Cultural Artifacts

K. Danziger, a Canadian psychologist, once spent two years teaching at a university in Indonesia[7]. Upon discovering that the host university already had a type of psychologist whose teachings were based on Hindu philosophy, Danziger suggested that the two of them organize a joint seminar in which Eastern and Western approaches to psychological problems could be discussed. However, they could not agree on what constituted a suitable topic of discussion. When Danziger suggested potential topics such as motivation, learning, or intelligence, his Indonesian colleague objected that the things Danziger included under each of these headings were heterogenous collections of phenomena that had nothing interesting in common. Conversely, the topics suggested by the Indonesian appeared incomprehensible to Danziger. The seminar never took place.

The difficulties experienced by Danziger and his Indonesian colleague suggest that the different mental processes recognised by conventional psychology are not natural kinds. In other words, they need not refer to real entities: they appear to be merely cultural artifacts invented by ancient philosophers. Danziger cites various other authors to show that Chinese and African societies each evolved their own distinctive systems of psychology. The reason that similar conventions are found in different Western countries is probably that the early Christian church adopted Aristotle's teachings and disseminated them widely over a period of many centuries. The doctrines of the great Catholic philosopher Saint Thomas Aquinas (c. 1225-1274) are largely based on those of Aristotle[8].

The fact that mentalistic terms are a useful part of our everyday speech may appear to provide an argument that such terms must refer to some real aspect of brain function. However, it must be understood that language relies very heavily on metaphors and many expressions that are nonsense from a biological point of view continue to be useful in ordinary speech and writing. We understand very well what is meant by such phrases as "bleeding hearts", "learning by heart", and so forth, even though we are well aware that the only known function of the heart is to pump blood through the vessels. Similarly, mentalistic terms need have no relation to the actual functioning of the brain.

IV. There is no Generally Valid Objective Criterion of Subjective Experience

It is generally accepted that subjective experience is wholly private and cannot be directly communicated to anyone else. There is no way to communicate the nature of the experience of red or the appearance of a starry night sky to someone who has been born blind. One of the consequences of this privacy of subjective experience is that there are no clear objective criteria by which the presence of subjective awareness can be determined by an outside observer. Although most people would agree that the thermostat that switches their home furnace on and off probably does not experience sensations of cold or warmth, opinions differ on whether, for example, fresh oysters served on the half shell suffer pain as they are eaten alive. Some scientists have suggested that even bacteria have a form of consciousness; others ridicule the idea.

The problem of determining whether subjective awareness is present or not becomes acute in some human accident victims in whom extensive injury to the brainstem removes all possibility of speech or movement. Some such patients, who have eventually recovered, state that they were quite aware of what was going on around them even though they were completely incapable of communicating with other people (locked-in syndrome). They can often report correctly events or conversations that took place during their period of unreactivity. An example of this type of phenomenon is provided by the case of Andrea Ostrum who was found in a ditch in a deep coma due to head injuries sustained in a car accident [9]. Dr. Ostrum eventually recovered the ability to speak, write (on a computer), eat, drink, and dress herself but she writes "One thing is certain: I was conscious long before I could let anyone know. I was aware of everything that was going on around me - I was unable to let anyone know because the accident left me paralysed below the neck, unable to signal in any way, and all the technology showed damage so extensive that nobody with that much damage could possible survive and be aware." In fact, Dr. Ostrum's coma appeared so profound that an application was begun to a court to ask for permission to remove her feeding tube and allow her to die. Fortunately, this was not done and eventually she recovered.

Consequently, there is reason to believe that subjective experiences may sometimes be present in a totally unreactive being. Whether the converse, i.e. the occurrence of waking intelligent behavior unaccompanied by subjective experience, ever occurs is unknown and may be forever unknowable. We cannot know, for example, whether a computer-controlled robot could ever be conscious. For this reason, it is difficult or impossible to test any proposal that some specific detail of neuronal structure or function is responsible for producing subjective awareness.

The difficulty is primarily one of identifying negative cases. It seems reasonable, though not open to proof, to believe that all normal intact adult humans have a capacity for subjective experiences but it is difficult or impossible to decide when this capacity has disappeared as a result of brain damage, drug effects, and so forth. Some surgical patients complain bitterly (and start law suits) on the grounds that they were fully conscious and in great pain during their surgery even though they appeared to the anesthetist to be fully anesthetized. However, even if the patients do not complain, we are not justified in assuming that they did not experience pain. Perhaps they were conscious but amnesic, that is, they experienced pain but do not remember it. If this possibility seems rather far-fetched, consider the situation occurring during normal physiological sleep. People ordinarily remember little or nothing of what occurred during a good night's sleep but if they are awakened during slow wave sleep (identified electroencephalographically) they usually claim to have been thinking about something and not really asleep at all. It seems to be possible for reduced reactivity, subjective awareness and an amnestic state to co-exist.

These limitations are of major importance in any attempt to relate objective features of brain activity to subjective experience. If we cannot determine whether subjective experience is present or not in specific cases all such efforts are doomed to failure.

V. The Application of Mentalistic Concepts to Brain Research Produces Bad Science

Research on the function of the hippocampal formation provides a clear example of the effect that unthinking acceptance of conventional psychological ideas can have on neuroscientific research. In the early twentieth century, neuroanatomists concluded that the dentate gyrus and Ammon's horn must have a predominantly olfactory function because they receive a strong olfactory input from the entorhinal cortex [10]. It came to be widely accepted that the small granule cells of the dentate gyrus, in particular, had a sensory function analogous to the sensory function of the small granule cells of layer 4 of the neocortex. Similarly, the large pyramidal cells of Ammon's horn were assumed to have an effectory or motor function much like the large pyramidal cells of layer 5 of the neocortex. Such suggestions are consistent with a scheme whereby the functional relations of the various areas of the entire cerebral cortex are classified in terms of their major sensory inputs [11]. Thus, the posterior neocortex is dominated by visual inputs, the anterior neocortex is dominated by somatosensory inputs, and the lateral neocortex is dominated by auditory inputs. According to this view, the entire pyriform lobe and hippocampal complex deserve the designation "rhinencephalon" because the activity of these cortical areas is dominated by olfactory inputs. Thus, this classification of the cerebral cortex is based fundamentally on anatomy, specifically on the central connections of the major sense organs.

In the mid twentieth century this anatomical view was challenged by a rival view which proposed that the function of certain cerebral cortical areas was specialized in relation to conventional psychological concepts. Perhaps the most influential of these suggestions was the proposal that the hippocampus is essential to certain types of memory. This view originated with W.B. Scoville and Brenda Milner of McGill University, who reported that large surgical removals of medial temporal lobe structures, including at least part of the hippocampal formation, produced a severe disorder of learning and memory in human patients. The hippocampal-memory concept dominated research and thinking on the hippocampal formation for more than forty years, inspiring hundreds of original papers, numerous research conferences and general acceptance by the writers of textbooks.

In retrospect, now that the idea that the hippocampus is essential for certain types of memory has generally been abandoned,[12] it may seem strange that it took so long to correct such a fundamental error. The whole affair of the function of the hippocampus raises questions about the validity of the conventional approach to research on the brain and behavior.

It is essential to achieve conceptual clarity. What are we trying to find out?

The conventional mentalistic approach to the brain-behavior field suggests that one should attempt to discover the neural basis of such processes as perception, attention, motivation, memory and so forth. Overt behavior is of interest only insofar as it serves as a means of identifying the presence of an underlying pattern of mental activity. Consequently, according to this approach, one should develop behavioral tests for the mental activity of interest such as the tests of memory (e.g. delayed match to sample, delayed non-match to sample) used so extensively in research on the hippocampus. The details of these tests are not important for the argument made here. What is important is that concern with mental processes leads to a neglect of behavior in a general sense. For example, in research on humans, it is quite astonishing that no one attempted to study the olfactory abilities of Scoville and Milner's famous patient H.M. until the mid-1980s, nearly three decades after the medial temporal lobe structures had been surgically removed [13]. G.A. Talland briefly described the inactivity and lack of spontaneous social behavior in patients with Korsakoff's syndrome in 1965 [14] but no one has attempted to explore this subject further. It seems likely that so-called amnestic syndrome patients suffer from a rather widespread impairment of behavior which has never been adequately studied. Rats with hippocampal lesions are defective, not only in so-called tests of memory and cognitive mapping, but also in instinctive behaviors such as food hoarding, nest building, maternal behavior, and body grooming. It is likely that humans with medial temporal lobe lesions would also be found to display widespread behavioral impairments if the subject were to be studied by adequate methods. Perhaps the amnestic syndrome should be viewed as a mild form of dementia, a syndrome in which there is no doubt that many behavioral functions are impaired.

A focus on memory, an unseen internal process, with a consequent neglect of behavior has also invalidated a large body of electrophysiological investigations of hippocampal activity during performance in learning and memory tasks. Training an animal does indeed produce changes in hippocampal electrical activity but the changes are due to the fact that training changes the amount of time an animal spends moving about or sniffing or because the body temperature has changed. This tells us nothing whatever about learning and memory [15].

What are learning and memory exactly? To an animal behaviorist the terms refer to experience-dependent adaptive changes in behavior occurring within the lifetime of one individual. A neuroscientist is likely to assume that the terms refer to long lasting experience-dependent changes in the efficacy of synaptic transmission in the central nervous system. To a psychologist the terms refer to mental processes distinct from those involved in perception, attention, motivation or emotion.

These differing assumptions suggest different types of experimental investigation. An experimental approach which has been popular with those who, implicitly or explicitly, adopt a psychological definition of memory has been to test the effect of surgical removal of a brain structure (such as the hippocampus) on performance in a so-called test of memory such as the delayed non-match to sample test. To someone who bases his work on the neuroscientific concept of memory, this method is far too crude to be of any value. From a neuroscientific point of view, it is obvious that any form of whole animal behavior will involve the operation of many billions of synapses. Function at some of these synapses may be modified by training but others, presumably the majority, will not be affected. Destruction of a large neural structure, such as the hippocampus, will produce a complex effect owing to the loss of neurons and neuronal connections that were not modified by training as well as those that were modified by training. It should be understood that studies of the effects of various brain lesions or drug treatments (which affect synapses in a transmitter-specific fashion regardless of whether they are plastic or not) on so-called tests of memory may teach us something about the neurology of a specific behavioural performance but have little direct relevance to the neuroscientific problem of synaptic plasticity.

The lack of progress in the attempt to apply mentalistic concepts to the function of the hippocampus is not an isolated phenomenon. As a further example, a huge research effort has also been devoted to the elucidation of the neural basis of attention [16]. The process of attention has been related to: (a) peripheral filtering of non-attended inputs; (b) the thalamic intralaminar nuclei and the brain stem reticular formation; (c) the hippocampus; (d) the cingulate cortex; (e) the parietal cortex; (f) the frontal cortex; (g) the cholinergic projections from the basal forebrain to the neocortex; (h) noradrenergic projections from the locus coeruleus to the cerebral cortex; (i) long term potentiation in the entorhinal projections to the dentate gyrus and Ammon's horn; (j) the pyriform cortex; and (k) a miscellaneous group of structures including the amygdala and globus pallidus. None of these ideas is widely accepted; none has been conclusively rejected. They merely fade away as people lose interest in them. Therefore, the cumulative growth of knowledge, so characteristic of natural science, has not occurred in research on the neural basis of attention. After many decades of failure in the search for a neural mechanism corresponding to the concept of attention, we must consider the possibility that it does not exist.

There are numerous other examples of how a preoccupation with the ancient theory of the psyche or mind has diverted investigators from a straight-forward study of the relation of cerebral activity to behavior and physiological processes (see Chapters 9 and 10). The apparent determination of many brain-behavior investigators to ignore behavior is truly rather remarkable. When giving talks at meetings, I have often met people who express astonishment and disbelief that such a trivial event (from their point of view) as walking a few steps should be associated with extensive changes in the activity of the hippocampus, neocortex and cingulate cortex even though the phenomena are obvious if one takes the trouble of looking for them.

VI. How Should we Approach the Brain-Behavior Problem?

I think it is time to recognise that the mentalistic approach to the study of brain function is not working. What is the alternative?

One of the great discoveries of twentieth century biology is that animal behavior can be profitably studied without regard to ancient doctrines concerning the psyche. If one thinks of behavior as a biological property of animals, comparable to digestion or circulation, it becomes apparent that: (a) possession of an extensive repertoire of adaptive behavior is of immense survival value; and (b) virtually all body systems in present day mammals have evolved to support a high level of behavioral output. For example, the skeleton and muscular systems of ancestral reptiles were modified extensively in mammals to permit free movement of the head and metabolically efficient locomotion.[17] Endothermy, and the high metabolic activity required for it, appear to have evolved in mammals (and in birds) because it makes possible sustained high levels of motor activity.[18] Reptiles, in comparison with mammals and birds fatigue very rapidly. Many aspects of the internal structure and function of mammals can be viewed as adaptations to permit the high metabolic rate required to sustain a continuous high level of motor activity. Thus in comparison with typical reptiles, mammals have evolved: (a) a muscular diaphragm to assist in gaseous exchange in the lungs; (b) complete separation of arterial and venous blood in the heart, together with a high blood pressure and high cardiac output; (c) a capacity for breaking food into small pieces and an overall improvement in the efficiency of digestion to permit the rapid processing of large amounts of food; and (d) adaptations such as fur, sweat glands, panting, etc., to regulate body temperature.

One can assume that the brain too will be organized in relation to the behavior it evolved to produce. As H.J. Jerison put it, "The brains of all species evolved in ways appropriate to control the behavior required for life in their environmental niches"[19]. In fact there is reason to think that the entire nervous system evolved to permit spontaneous motor activity and prompt reactivity to environmental stimuli. An interesting illustration of this is provided by the life history of the tunicates (subphylum *Urochordata*, also known as ascidians or sea squirts). A larval sea squirt somewhat resembles a tadpole. It has a notochord (an evolutionary precursor of the vertebral column), a hollow dorsal nerve cord, and a pharynx containing gill slits. After a period of free-swimming existence, the larva attaches itself to a rock or wooden wharf and undergoes a transformation to a 2-3 cm long bag-like creature that spends the rest of its life filtering sea water to extract the minute organic food particles it contains. The notochord disappears and the dorsal nerve cord is reduced to a small ganglion.[20] Thus, creatures with an active life seem to require a nervous system but those whose life is entirely passive and vegetative can largely or entirely dispense with it.

The observations discussed in this book support the view that the large-scale features of brain activity are organized in close relation to overt behavior and sensory inputs. Consequently it may be possible to replace the apparently unsolvable three topic problem of the brain, mind, and behavior by a more tractable two topic problem of brain and behavior.

A critical reader may well conclude at this point that a preoccupation with sensory and motor processes ignores many of the most interesting topics in the brain-mind-behavior field. What, for example, could such an approach possibly have to say about self-awareness, something regarded by many people as a uniquely human attribute?

I once published a paper that is relevant to the question of self-awareness in the rat.[21] About 1970-71, I was much interested in the similarity of hippocampal rhythmical slow activity during active sleep (rapid eye movement sleep) and during running or jumping in the waking state. Evidence available at the time indicated that brain motor systems are in vigorous activity during active sleep but that muscular expression of this activity is strongly suppressed by inhibition of spinal motor neurons and reflex afferents. Perhaps rats dream about running and jumping. Would it be possible to train them to tell us?

Steve Kendall, a colleague in the Psychology Department, was actively engaged in research on operant conditioning at that time. We both knew a talented undergraduate student, Richard (Rick) Beninger, who was interested both in brain-behavior research and in operant conditioning. I proposed to hire Rick as a research assistant during the summer of 1972 to allow him to train rats to report on their own behaviour using the operant conditioning equipment available in Steve's lab. The idea was to train rats to respond to a signal, X, by pressing one lever if they were engaged in behaviour A, a different lever if they were engaged in behavior B, and so forth. After this preliminary training, electrodes could be implanted and the signal X could be presented during different phases of sleep. One can think of the presentation of X as a question: "What are you doing?" The rat can then reply by pressing an appropriate lever.

The procedure that the three of us developed was to train rats in a large semi-circular arena containing up to 4 different levers. Whenever a buzzer sounded, pressing one, and only one, of the 4 levers resulted in the presentation of 0.1 ml of sweetened condensed milk, a very attractive food for rats. The correct lever varied randomly from trial to trial but depended on the behaviour in which the rat was engaged when the buzzer sounded. If the rat was immobile, lever #1 was correct; if the rat was walking, lever #2 was correct; if the rat was washing its face, lever #3 was correct; and if the rat was rearing on its hind legs, lever #4 was correct. Great care was taken to ensure that the discrimination was not based on unintended environmental cues. The experimenter (Rick) controlled the experiment remotely from an adjacent room while observing the rat with the aid of a video camera.

Although the rats became very proficient at discriminating their own behaviour, the training procedure occupied the entire summer. Since the beginning of classes in the fall term made continued work impossible, the second phase of the experiment, which would have involved implanting electrodes and testing the effect of sounding the buzzer during different sleep states, was never carried out. Rick, Steve and I published the results we had as a demonstration that rats can discriminate and report on their own behaviour.

A rat could also be trained to press one lever when a white triangle is presented, a second lever when a white circle is presented, and so forth. In both cases we have examples of operant behaviour under discriminative stimulus control. In one case the stimulus control is exerted by visual inputs; in the other case it is exerted, probably, by proprioceptive inputs. There is a considerable experimental literature showing that human self-report is largely based on proprioceptive and enteroceptive inputs (see reference in Note #1). One can well imagine that the behavioral phenomena of operant conditioning under stimulus control could be demonstrated by a mindless computer-operated robot. Self-report can be regarded purely as a behavioral ability. There is at present no known way of deciding whether or not such an ability is associated with a subjective experience of the self.

The approach to the brain and behaviour which I advocate is restricted in scope since it focuses exclusively on objective facts, refuses to make inferences based on conventional psychological concepts, and attempts to account for the occurrence of behavior entirely in terms of the activity of neurons. Speaking as someone who has been a laboratory worker for more than 40 years, my preference is to work on potentially soluble problems, even though their scope may be somewhat limited, rather than engage in a fruitless pursuit of grand philosophical questions whose answers may lie forever beyond our reach. The ultimate nature of subjective awareness may be similar, in terms of its degree of incomprehensibility, to such topics as the ultimate nature of matter (What, exactly, *is* an electron, or a quark?), and of time, space, force, causality, or more simply, the question of the existence and nature of life in galaxies other than the Milky Way. We live surrounded by the unknown and the unknowable, like primitive hunters crouching at night around a tiny fire whose flickering light allows them to see only a short way into the vast cold forest that surrounds them. Although this is not a comforting thought, it appears to correspond to reality.

Notes on Chapter XI

1. Vanderwolf, C.H. (1998). Brain, behavior, and mind: What do we know and what can we know? *Neuroscience and Biobehavioral Reviews, 22*: 125-142.
2. Newton, Sir Isaac. (1931). *Opticks*. London: G. Bell & Sons Ltd., 1931 (Reprinted from the IVth edition, 1730), pp 124-125.
3. Haldane, E.S. and Ross, G.R.T. (1955). *The philosophical works of Descartes*: *Volume I*. New York: Dover Publications, (First published by Cambridge University Press, 1911, reprinted with corrections, 1931). The lines quoted are on pp 145-146, and p. 157.
4. Barnes, J. (ed) (1984). *The complete works of Aristotle, volumes 1 & 2*. Princeton, N.J., Princeton University Press.
5. On the history of introspective psychology see:
   a) Boring, E.G. (1953). A history of introspection. *Psychological Bulletin, 50*: 169-189.
   b) Hebb, D.O. (1980). *Essay on mind*. Hillsdale, N.J.: Lawrence Erlbaum.
   c) Humphrey, G. (1951). *Thinking: an introduction to its experimental psychology*. New York: Wiley.

   On the absence of direct knowledge of the causes of our own behavior see:
   a) Gopnik, A. (1993). How we know our minds: the illusion of first person knowledge of intentionality. *The Behavioral and Brain Sciences, 16*:1-14.
   b) Hebb, D.O. (1977). To know your own mind. In Nicholas, J.M. (ed.) *Images, perception and knowledge*. Dordrecht: Reidel; 213-219.
   c) Lyons, W. (1986). *The disappearance of introspection*. Cambridge, Mass., MIT Press.
   d) Nisbett, R.E., and Wilson, T.D. (1977). Telling more than we can know: verbal reports on mental processes. *Psychological Review, 84*: 231-259.
6. Lashley, K.S. (1958). Cerebral organization and behavior. *Proceedings of the Association for Research in Nervous and Mental Disease, 36*: 1-18.
7. Danziger, K. (1997). *Naming the mind: how psychology found its language*. London: Sage Publications.
8. Russell, B. (1961). *History of western philosophy*. London: Allen and Unwin.
9. Ostrum, A.E. (1994). The 'locked-in' syndrome - comments from a survivor. *Brain Injury, 8*: 95-98.

10. A summary of evidence indicating that the hippocampal formation plays a role in the olfactory control of behavior is provided in Chapter 10 and by: Vanderwolf, C.H. (2001). The hippocampus as an olfacto-motor mechanism: were the classical anatomists right after all? *Behavioural Brain Research, 127*: 25-47.
11. Diamond, I.T. (1985). A history of the study of the cortex: changes in the concept of the sensory pathway. In: G.A. Kimble and K. Schlesinger (eds.) *Topics in the history of psychology,* Hillsdale, New Jersey: Lawrence Erlbaum Associates, pp. 305-387.
12. Gaffan, D. (2001). What is a memory system? Horel's *critique* revisited. *Behavioural Brain Research, 127*: 5-11.
Horel, J.A. (1978). The neuroanatomy of amnesia: a critique of the hippocampal memory hypothesis. *Brain, 101*: 403-445.
Horel, J.A. (1994). Some comments on the special cognitive functions claimed for the hippocampus. *Cortex, 30*: 269-280.
Vanderwolf, C.H. and Cain, D.P. (1994). The behavioral neurobiology of learning and memory: a conceptual reorientation. *Brain Research Reviews, 19*: 264-297.
13. Eichenbaum, H., Morton, T.H., Potter, H. and Corkin, S. (1983). Selective olfactory deficits in case H.M. *Brain, 106*: 459-472.
14. Talland, G.A. (1965). *Deranged memory: a psychonomic study of the amnesic syndrome*, New York: Academic Press.
15. This issue is discussed more extensively in the papers by Vanderwolf (2001) (see Note #10) and by Vanderwolf and Cain (see Note #12).
16. A sampling of references to studies of the neural basis of attention:
    a) Peripheral filtering of non-attended stimuli: Hernández-Peón, R., Scherrer, H., and Jouvet, M. (1956). Modification of electrical activity in cochlear nucleus during "attention" in unanesthetized cats. *Science, 123*: 331-332.
    b) Thalamic intralaminar nuclei and brain stem reticular formation: Jasper, H.H.(1960). Unspecific thalamocortical relations. In J. Field, H.W. Magoun and V.E. Hall (eds). *Handbook of physiology, Section 1: Neurophysiology, volume 2.* Washington, D.C. American Physiological Society, pp. 1307-1321.
    Lindsley, D.B. (1960). Attention, consciousness, sleep and wakefulness. In J. Field, H.W. Magoun, and V.E. Hall (eds) *Handbook of physiology, Section 1: Neurophysiology, volume 3.* Washington, D.C. American Physiological Society, pp. 1553-1593.
    c) The hippocampus:
    Bennett, T.L. (1975). The electrical activity of the hippocampus

and processes of attention. In R.L. Isaacson and K.H. Pribram (eds). *The hippocampus, volume 2: Neurophysiology and behavior*. New York: Plenum Press, pp. 71-99.

d) The cingulate cortex:
Kaada, B.R. (1960). Cingulate, posterior orbital, anterior insular and temporal pole cortex. In J. Field, H.W. Magoun, and V.E. Hall (eds). *Handbook of physiology, Section 1: Neurophysiology, volume 2*. Washington, D.C. American Physiological Society, pp. 1345-1372.

e) Parietal and frontal cortex:
Colby, C.L., and Goldberg, M.E. (1999). Space and attention in parietal cortex. *Annual Review of Neuroscience, 22*: 319-349.
Kolb, B. and Whishaw, I.Q. (2001). *An introduction to brain and behavior*. New York: Worth Publishers, (see pp 537-539).

f) Cholinergic projections from the basal forebrain:
McGaughy, J., Everitt, B.J., Robbins, T.W., and Sarter, M. (2000). The role of cortical cholinergic afferent projections in cognition: impact of new selective immunotoxins. *Behavioural Brain Research, 115*: 251-263.

g) Ascending noradrenergic projections:
Aston-Jones, G., and Bloom, F.E. (1981). Activity of norepinephrine-containing locus coeruleus neurons in behaving rats anticipates fluctuation in the sleep-waking cycle. *Journal of Neuroscience, 1*: 876-886.
Mason, S.T., and Iverson, S.D. (1979). Theories of the dorsal bundle extinction effect. *Brain Research Reviews, 1*: 107-137.

h) Hippocampal long-term potentiation:
Shors, T.J., and Matzel, L.D. (1997). Long-term potentiation: What's learning got to do with it? *The Behavioral and Brain Sciences, 20*: 597-655.

i) Amygdala:
Gloor, P. (1960). Amygdala. In J. Field, H.W. Magoun, and V.E. Hall (eds). *Handbook of physiology, Section 1: Neurophysiology, volume 2*. Washington, D.C. American Physiological Society, pp. 1395-1420.

j) Pyriform cortex:
Freeman, W.J., and Skarda, C.A. (1985). Spatial EEG patterns, non-linear dynamics and perception: the neo-Sherringtonian view. *Brain Research Reviews, 10*: 147-175.

17. Radinski, L.B. (1987). *The evolution of vertebrate design*. Chicago: University of Chicago Press.

18. Bennett, A.F., and Ruben, J.A. (1979). Endothermy and activity in vertebrates. *Science, 206*: 649-654.
19. Jerison, H.J. (1973). *Evolution of the brain and intelligence.* New York: Academic Press, (see p.15).
20. Weichert, C.K. (1970). *Anatomy of the chordates, 4th edition.* New York: McGraw-Hill Book Company.
21. Beninger, R.J., Kendall, S.B., and Vanderwolf, C.H. (1974). The ability of rats to discriminate their own behaviours. *Canadian Journal of Psychology, 28*: 79-91. Rick Beninger went on to do graduate work at McGill University with Peter Milner and eventually became a professor in two departments, Psychology and Psychiatry, at Queen's University in Kingston, Ontario.

Chapter 12

Epilogue

I have enjoyed my years in the lab. It is a marvellous thing to discover some new phenomenon that no one has ever seen before or to understand something that no one has ever understood before. New phenomena have been especially appealing to me, in part because they are intrinsically fascinating, rather like finding a new and strangely patterned shell on a sandy beach, and also because they are beyond dispute. Critics and doubters can be silenced by a simple demonstration.

Most of the work done by my students and me can be viewed as a series of attempts to analyse phenomena which were themselves discovered largely by chance. Although hypothesis testing was a part of the overall endeavour, it has not been a primary aim of the research. I have very little faith in extensive *a priori* hypotheses, perhaps because most of mine turned out to be wrong or irrelevant. The account given in this book under-represents the extent of this because the only ideas that are discussed are those that led to some type of successful conclusion. During my career I carried out an enormous number of one- or two-rat experiments that were promptly abandoned. There is nothing like an experiment to demonstrate that many of the novel ideas that one has are not even wrong but merely ridiculous.

I hope that the demonstration that many aspects of the field potential activity of the cerebral cortex are closely related to behavior will help to interest neuroscientists in objective behavioral phenomena. There is a great need of this not only in neuroscience but also in related fields. Psychiatry, for example, has been strongly influenced by a point of view which has been well expressed by K.

Jaspers who wrote: "Psychopathology has *as its subject-matter*, actual conscious psychic events." (italics in the original)[1]. Needless to say, careful observation of behavior is of little importance to those who accept this approach. The following example illustrates this.

During the period that I worked on serotonin, I became interested in depression, a common human affliction that may often be due to dysfunction of brain serotonergic mechanisms. However, reading papers in the field did not provide me with a clear idea of what depression is. (I had never actually seen a depressed patient.) There was much discussion of "negative affect" and so forth, but I could not find a systematic description of what depressed patients actually do. Eventually, I discovered investigations of depression[2] that made use of observational methods derived from the study of animal behavior. Depression, it seems, is characterised by general inactivity and a lack of social interaction but may be associated with endless weeping and stereotyped pathological behavior. In rats, serotonin release in the forebrain seems to promote Type 1 behavior.[3] Perhaps depressed patients are deficient in the cerebral control of the human equivalent of rat Type 1 behavior (resulting in low levels of locomotion, manipulation and speaking) allowing pathological Type 2 behaviors (weeping, hand-wringing) to occur frequently. Although such a hypothesis is certain to be inadequate (or even ridiculous) it may suggest novel approaches to the study of a major psychiatric illness.

<u>Notes on Chapter 12</u>

1. Jaspers, K. (1963). *General psychopathology.* Manchester, U.K. Manchester University Press.
2. Fossi, L., Faravelli, C., and Paoli, M. (1984). The ethological approach to the assessment of depressive disorders. *Journal of Nervous and Mental Disease, 172*: 332-341.
   Pedersen, J., Schelde, J.T.M., Hannibal, E., Behnke, K., Nielsen, B.M., and Hertz, M. (1988). An ethological description of depression. *Acta psychiatrica scandinavica, 78*: 320-330.
3. Takahashi, H., Takada, Y., Nagai, N., Urano, T., and Takada, A. (2000). Serotonergic neurons projecting to hippocampus activate locomotion. *Brain Research, 869*: 194-202.
   Vanderwolf, C.H., McLauchlin, M. Dringenberg, H. C. and Baker, G.B. (1997). Brain structures involved in the behavioral stimulant effect of central serotonin release. *Brain Research, 772*: 121-134.

Index